AF811641

Chemically Modified Minds

Matthew Hall
Mark Forshaw • Catharine Montgomery
Editors

Chemically Modified Minds

Substance Use for Cognitive Enhancement

palgrave
macmillan

Editors
Matthew Hall
Arden University
Coventry, UK

University of Derby
Derby, UK

Mark Forshaw
Liverpool John Moores University
Liverpool, UK

Catharine Montgomery
Liverpool John Moores University
Liverpool, UK

ISBN 978-981-15-6770-4 ISBN 978-981-15-6771-1 (eBook)
https://doi.org/10.1007/978-981-15-6771-1

© The Editor(s) (if applicable) and The Author(s) 2021
This work is subject to copyright. All rights are solely and exclusively licensed by the Publisher, whether the whole or part of the material is concerned, specifically the rights of translation, reprinting, reuse of illustrations, recitation, broadcasting, reproduction on microfilms or in any other physical way, and transmission or information storage and retrieval, electronic adaptation, computer software, or by similar or dissimilar methodology now known or hereafter developed.
The use of general descriptive names, registered names, trademarks, service marks, etc. in this publication does not imply, even in the absence of a specific statement, that such names are exempt from the relevant protective laws and regulations and therefore free for general use.
The publisher, the authors and the editors are safe to assume that the advice and information in this book are believed to be true and accurate at the date of publication. Neither the publisher nor the authors or the editors give a warranty, expressed or implied, with respect to the material contained herein or for any errors or omissions that may have been made. The publisher remains neutral with regard to jurisdictional claims in published maps and institutional affiliations.

Cover illustration: © palau83

This Palgrave Macmillan imprint is published by the registered company Springer Nature Singapore Pte Ltd.
The registered company address is: 152 Beach Road, #21-01/04 Gateway East, Singapore 189721, Singapore

Acknowledgements

Editing a collection of divergent research which actively evaluates the emergence of, and apparent increasing use of, non-prescription and prescription drugs for use in enhancing cognition is not always an easy task. However, this challenge has been made all the more manageable, thanks to the flexibility, sensitivity and collegiate spirit demonstrated by the contributors in this volume. Collectively, we would like to thank them for their contributions and hope we have done them justice in conveying their voices and their perspectives.

We would also like to thank Joshua Pitt at Palgrave Macmillan, who waited patiently for the manuscript and who saw it through to publication, and the anonymous reviewers for their feedback.

Matthew Hall, Mark Forshaw & Catharine Montgomery, 2020

Contents

Notes on Contributors

Sarah Benson is a postdoctoral research fellow at the Centre for Human Psychopharmacology, Swinburne University. She completed her PhD in 2016, where she explored the cognitive and behavioural effects of alcohol and caffeine. Her research portfolio has focused primarily on cognitive impairment and enhancement, alcohol hangover and drug use and functional outcomes. Benson aims to disentangle the neurocognitive effects of licit and illicit substances, to identify relationships between their behavioural effects and physiological properties and to reduce dangerous behaviours and harm.Benson has been awarded over $2 million in research funding and over 30 peer-reviewed journal articles and several book chapters of Benson have been published. She is committed to science communication and has appeared numerous times in television, radio, online and print media.

Vince Cakic is a research assistant with the University of Sydney investigating the similarities and departures between psychosis and mystical experience, the hypothesis being that they are the same construct under two antipodal names. He holds a BS (Psychol) degree. In his spare time, he is examining a novel zero-day vulnerability affecting one of the largest vendors of electronic medical record (eMR) software in the world.

Robert C. Dempsey is a senior lecturer in the Department of Psychology at Manchester Metropolitan University. A common theme to Dempsey's research is the role of individual perceptions of the social environment and their influences on mental and physical health-related outcomes. His work focuses on three areas: (1) suicidality, (2) bipolar disorder and (3) evaluating the role of social norms perceptions in health-related behaviour. Dempsey is particularly interested in understanding how social influence factors and perceived norms influence individuals' intentions to engage in a range of positive and negative health behaviours, and how information interventions which challenge commonly held under/over-estimations of social norms can encourage more healthy behaviours. He has previously worked on a European Commission-funded feasibility trial of a social norms feedback intervention for polysubstance use in university students and is working on various projects applying and evaluating this approach with other health behaviours.

Jade M. Elliott is a senior lecturer at Staffordshire University (UK). She is a cognitive/biological psychologist with primary interests in the neurological dysfunction and decline associated declining glucoregulatory control, as found in normal ageing and with the onset of diabetes. In particular, Elliott focuses on the nuances of memory and cognitive changes in preclinical populations with risk factors associated with declining glucoregulatory control.

Mark Forshaw is a Health Psychologist and Subject Leader at Liverpool John Moores University. He is a Past President of the Institute of Health Promotion & Education, and was a trustee of the British Psychological Society amongst other organisations. He is the author of numerous books on health psychology, research, and critical thinking.

Matthew Hall is the Programme Leader for Psychology, and Chair of the Research and Ethics Committee at Arden University. He is also Associate Academic, University of Derby and Editor of the *Journal of Gender Studies*. He is a prolific writer and his research interests include health; new, mediated and cyberspace identities; image-based sexual violence; online violations; cognitive enhancement; body modification; disability; gender and sexuality; body image and image consciousness.

Catharine Montgomery is Reader in Psychopharmacology at Liverpool John Moores University (LJMU). She is co-leader of the Research Centre for Brain and Behaviour at LJMU, and is on the steering group for

Liverpool Centre for Alcohol Research. She has expertise in the effects of substance use on psychological functioning, and has authored numerous journal articles and book chapters on this subject.

Steven Roodenrys is a cognitive psychologist at the University of Wollongong, Australia. He has a background in research on short-term and working memory which includes work on basic processes in adults and children, and on working memory deficits in children with attention deficit hyperactivity disorder (ADHD) and reading disability. He has been involved in research on improving cognitive function, through cognitive training in children with ADHD, and a number of randomised trials looking at the impact of nutritional interventions on cognitive function in adults, including adults with dementia. These studies have involved supplements and whole foods.

Andrew Scholey is director of the Centre for Human Psychopharmacology at Swinburne University, Melbourne. He has led more than hundred studies in experimental psychopharmacology including into the neurocognitive effects of nutritional interventions (including whole diet and food components) metabolic substrates, recreational drugs, alcohol and hangover. Scholey has written upwards of 250 papers, been awarded more than $25 million in research funding. He has advised the UK and Canadian government on health policy. His current research focuses on understanding the mechanisms of cognitive impairment, enhancement and neuroprotection. Scholey works closely with industry to allow rapid translation of research into evidence-based, end-user benefits to brain health.

Jamie L. Tully is a recent graduate from Liverpool John Moores University. Tully completed his PhD in psychopharmacology investigating the use of various cognition-enhancing drugs among students at UK universities. He is particularly interested in the novel pharmaceutical stimulant modafinil and the long-term effects the substance exerts on executive functioning and neurophysiological response. His experience also includes using the neuroimaging modality functional Near-Infrared Spectroscopy to examine neurovascular indices of underlying cognitive deficits in modafinil users.

1

Introduction

Jamie L. Tully, Mark Forshaw, Matthew Hall, and Catharine Montgomery

What Are Cognitive Enhancement Drugs?

The most popular CE drugs belong to two groups: soft enhancers and PCE. Soft enhancers are popular, legally available substances which include food products, herbals substances and tonics, and products containing caffeine. Conversely, PCE use is often prohibited and includes synthetic pharmaceutical substances and some illegal drugs (Maier, Ferris, & Winstock, 2018). Both CE groups differ in the magnitude of their effect on cognitive performance and in mechanism of action, although

J. L. Tully • M. Forshaw • C. Montgomery
Liverpool John Moores University, Liverpool, UK
e-mail: J.Tully@2010.ljmu.ac.uk; M.J.Forshaw@ljmu.ac.uk; C.A.Montgomery@ljmu.ac.uk

M. Hall (✉)
Arden University, Coventry, UK

University of Derby, Derby, UK
e-mail: mhall@arden.ac.uk

© The Author(s) 2021
M. Hall et al. (eds.), *Chemically Modified Minds*,
https://doi.org/10.1007/978-981-15-6771-1_1

variations also exist within each group. Differences between the various CE categories and individual substances are discussed below.

Soft Enhancement

Some soft enhancers are commonplace in society, particularly those products containing caffeine, which is the most widely consumed psychoactive substance in the world (Zhang, Jiang, Liu, & He, 2017). Caffeinated beverages are among the most popular caffeine-based products because they act as a minor stimulant and promote feelings of alertness and wakefulness, which often become integrated into a person's daily routine. Consequently, it has been suggested that use of caffeine for CE is less explicit than other soft enhancers or PCE (Rosen & Weil, 2004). For example, nutraceuticals (such as ginseng, Ginkgo biloba and bacopa monnieri) are more explicitly marketed as CE drugs but are significantly less popular, owing to the fact that these drugs each have medicinal properties that are reported to act differently on cognitive performance and mood, each with varying degrees of success (Rai, Bhatia, Sen, & Palit, 2003; Tsai, Lin, Simon Pickard, Tsai, & Mahady, 2012). Moreover, these substances are often available only at speciality retailers, which suggests that they are purchased with a specific intention in mind. Of course, nutraceutical drugs are not just marketed as CE substances; they are also lauded for their various physical health benefits (Ward et al., 2019), meaning the reasons for their consumption are not always clear.

Caffeine, a psychostimulant, acts on the autonomic nervous system and shifts dominance from the sympathetic nervous system (SNS) to the parasympathetic nervous system (PNS). Unlike PCE, its effects are non-selective as it acts through blocking adenosine receptors (A_1 and A_2) in the prefrontal cortex, which in turn promotes monoamine release. As such, the neurotransmitters dopamine, serotonin and noradrenaline are released, which increase feelings of wakefulness and alertness (Fredholm, Yang, & Wang, 2017). Therefore, caffeine shares more similarities with pharmaceuticals used for PCE than nutraceuticals which operate through enzyme synthesis (Ahmed et al., 2016), although effects are less pronounced and the substance has a shorter half-life than PCE (Franke et al.,

2017). Notable soft enhancers which contain caffeine include coffee, energy drinks, caffeine pills and guarana (Maier, Liakoni, Schildmann, Schaub, & Liechti, 2015).

Pharmacological Cognitive Enhancement

Distinct from soft enhancers, PCEs are typically synthetic pharmaceutical substances and sometimes illegal drugs, whose use is controlled or prohibited by law. Most studies on PCE effects focus on amphetamines, in particular dextroamphetamine (d-amphetamine), which is sold under the trade name 'Adderall', and racetams like piracetam. However, evidence in support of these drugs as effective PCE is limited, with the studies that exist demonstrating only modest enhancements with single periods of use. Reports suggest that the effectiveness of these drugs often does not meet or exceed user expectations (Bagot & Kaminer, 2014; Battleday & Brem, 2016; Linssen, Sambeth, Vuurman, & Riedel, 2014). Research has also looked at use of illegal drugs as PCE in the UK such as psychedelics (Elsey, 2017) and cannabis (Franke, Roser, Lieb, Vollmann, & Schildmann, 2016), but as use of these substances as CE appears to be rare they will not be discussed further. Instead, the focus will be on substances which are most commonly self-reported by users for PCE, namely d-amphetamine, methylphenidate (MPH) and modafinil.

Amphetamine is a central nervous system (CNS) stimulant which is used to treat attention deficit hyperactivity disorder (ADHD). D-amphetamine in particular has been found to modulate neurotransmitter networks, predominantly dopamine, serotonin and noradrenaline (Darracq, Blanc, Glowinski, & Tassin, 1998). Alterations to monoamine neurotransmission are linked to feelings of increased wakefulness and alertness in humans, which has also made the substance useful for the treatment of narcolepsy syndrome in the past (Parkes & Fenton, 1973). Furthermore, in adolescents with ADHD, studies show improved learning outcomes during schooling, owing to a long half-life, which extends the overall effect of the drug throughout the school day (Pelham et al., 1999). Moreover, dopaminergic drugs like Adderall are shown to positively impact mood, which evidence suggests can increase creative thought processes (Farah, Haimm, Sankoorikal, Smith, & Chatterjee, 2009). Benefits to cognition in healthy people do not appear to be as extensive as with other PCE, although findings are comparatively limited (Bagot & Kaminer, 2014; Ilieva, Boland, & Farah, 2013). However, users regularly self-report pleasurable experiences with the drug which may contribute to the perception of enhanced cognition (Vargo & Petróczi, 2016; Vrecko, 2013).

Similar to Adderall, MPH is prescribed for ADHD and sometimes narcolepsy. Furthermore, the pharmacology of both substances is similar, as MPH also modulates noradrenaline and dopamine in the prefrontal cortex which is linked to a reduction of symptoms associated with ADHD, as well as selective improvements in cognitive functioning (Linssen et al., 2014). In particular, MPH has been shown to benefit cognition in adolescent and adult ADHD sufferers, who exhibit improved reaction time, attention and executive and non-executive memory (Coghill et al., 2014; Storebø et al., 2015). Emerging evidence has also suggested that MPH can improve working memory deficits found in stimulant users, although more research must be conducted in the area to confirm these findings (Moeller et al., 2014). With healthy people, studies using MPH to improve cognitive functioning show less compelling results, though it can be beneficial to certain cognitive functions, including processing speed, inhibitory control, working memory and memory consolidation (Linssen et al., 2014). While ADHD sufferers gain the most benefits from MPH, self-administration by healthy adults has risen, possibly to achieve CE, although reasons behind use have not been fully explored (White, Becker-Blease, & Grace-Bishop, 2006). Nevertheless, this kind of use appears to be relatively safe, as the potential for harm with MPH is seen to be low when taken in clinically safe doses, with the most extreme and commonly reported side effects being appetite suppression and disturbed sleep, which are symptoms frequently associated with stimulant use (Becker, Froehlich, & Epstein, 2016; Jeffers & Benotsch, 2016).

Finally, modafinil is a novel psychostimulant which is primarily used to treat narcolepsy, but is also used for shift work sleep disorder. Modafinil is a dopamine and noradrenaline reuptake inhibitor, as well as having modulatory effects on histamine in the prefrontal cortex (d'Angelo,

Savulich, & Sahakian, 2017). Despite similarities with MPH and Adderall, modafinil has the longest half-life of the PCEs, lasting approximately 10–15 hours, which is comparatively longer than Adderall at 3.5 hours (Robertson & Hellriegel, 2003). Modafinil is a prescription drug in the UK, and is listed as a controlled substance in the US, restricting its possession by healthy people. However, because it is shown to increase feelings of alertness and wakefulness, research reveals that it has gained popularity as a CE drug. Moreover, healthy people also exhibit benefits to attentional processes, learning and memory (Sahakian et al., 2015; Turner et al., 2003) and self-report feelings of increased energy and alertness (Stoops, Lile, Fillmore, Glaser, & Rush, 2005). Similar to MPH, it appears to have low potential to cause harm when taken in clinical quantities, however; evidence of doses which exceed clinical guidelines is limited, and thus potential adverse events cannot properly be evaluated (Battleday & Brem, 2016; Rush, Kelly, Hays, Baker, & Wooten, 2002).

Prevalence Estimates

Recent studies have estimated the prevalence of CE substance use. The evidence indicates that use of CE substances is increasing in both the US (Advokat & Scheithauer, 2013; Emanuel et al., 2013) and Europe (Maier, Liechti, Herzig, & Schaub, 2013). However, robust and comparable estimates of use are difficult to obtain due to how data is collected. Primarily, survey questions which investigate use often lack standardisation, and drugs are not uniformly assessed between studies. Interpreting use is also a challenge, as many prevalence figures do not reveal the reasons behind use (e.g. CE or recreational), which can be further confused by cultural differences in what are considered CE drugs. There is also a lack of racial diversity in sample characteristics making generalisability of results between populations challenging, and the limited follow-up studies make it difficult to perform trend analyses.

Limitations notwithstanding, it is clear that use of soft enhancers, particularly coffee and other caffeinated products, is considerably greater than PCE (Maier et al., 2013; Singh, Bard, & Jackson, 2014; Wolff, Brand, Baumgarten, Lösel, & Ziegler, 2017). Common media perception is that students and those in high-pressure jobs may be more likely to use PCE. This is partially supported by anecdotal evidence in universities. For example, the University of Cambridge student newspaper surveyed 1000 Cambridge University students and found that 10% reported use of a stimulant-based cognitive enhancer (modafinil, methylphenidate or dextroamphetamine). While there have been no large-scale empirical studies in the UK to corroborate this, these levels seem broadly on a par with other universities in the rest of the world. In one New Zealand university, 6.6% of respondents had used a cognitive enhancer (Ram et al., 2016), while a multisite study in the US found comparable estimates of 6.9% (McCabe et al., 2005). Australia had a slightly higher prevalence of amphetamine use for cognitive enhancement at 10.9% (Mazanov, Dunn, Connor, & Fielding, 2013). In Europe, estimates of prevalence range from 5% to 46% (Schelle et al., 2015). One study in Switzerland found that 14% of participants reported nonmedical use of cognitive enhancers across their lifetime, though this was lower for past-year and past-month use (Maier et al., 2013), while in Germany estimates ranged from <1% to 20% (Dietz et al., 2013). The UK appears to have lower rates at <5% (Holloway & Bennett, 2012). However, a recent study and the biggest exploration of international CE use to date found that between 2015 and 2017, self-reported use of PCEs increased from 1.7% to 5.1%. Further still, use of modafinil in the UK was highest among the 15 countries surveyed, and saw a substantial increase of 3.2% in 2015 to 10% in 2017, which is consistent with previous claims in the media (Maier et al., 2018).

Outside the student population, it has been speculated that university professors use cognitive enhancers to allow them to keep up with the demands of academia (Sahakian & Morein-Zamir, 2011). Moreover, a survey of German surgeons found that 8.9% reported ever having used a substance for cognitive enhancement. Across both of these studies, reasons reported for using enhancers included improving mood, promoting wakefulness, counteracting jetlag and enhancing productivity. Conversely, students seem to report using substances more during periods of high stress, such as exam months (Maier, Liechti, Herzig, & Schaub, 2013) which could affect overall prevalence estimates.

Conclusions

In summary, cognitive enhancement drugs belong to two groups—soft enhancers and PCE. With PCEs, experimental studies show modest results, but acute benefits to some aspects of executive functioning such as working memory and attention have been observed. Furthermore, based on prevalence estimates in the UK, modafinil appears to be the most popular PCE, with recent data even showing a considerable increase in use. Nonetheless, very little robust peer-reviewed data exists which supports these claims. Problems with methodology and sample diversity also raise the issue of generalisability, and intentions behind nonmedical use are rarely investigated making it difficult to interpret prevalence data. Consequently this book has three broad themes: the uses and effects of nutraceuticals, the uses and effects of pharmacological enhancers and ethical considerations in pharmacological cognitive enhancement. Each chapter is intended to be self-contained and starts with an overview before the main topic is discussed in depth.

References

Advokat, C. D., & Scheithauer, M. (2013). Attention-deficit hyperactivity disorder (ADHD) stimulant medications as cognitive enhancers. *Frontiers in Neuroscience, 7*, 82.

Ahmed, T., Raza, S. H., Maryam, A., Setzer, W. N., Braidy, N., Nabavi, S. F., … Nabavi, S. M. (2016). Ginsenoside Rb1 as a neuroprotective agent: A review. *Brain Research Bulletin, 125*, 30–43. https://doi.org/10.1016/j.brainresbull.2016.04.002

Bagot, K. S., & Kaminer, Y. (2014). Efficacy of stimulants for cognitive enhancement in non-attention deficit hyperactivity disorder youth: A systematic review. *Addiction (Abingdon, England), 109*(4), 547–557.

Battleday, R., & Brem, A. (2016). Modafinil for cognitive neuroenhancement in healthy non-sleep-deprived subjects: A systematic review. *European Neuropsychopharmacology: The Journal of the European College of Neuropsychopharmacology, 26*(2), 391.

Becker, S. P., Froehlich, T. E., & Epstein, J. N. (2016). Effects of methylphenidate on sleep functioning in children with attention-deficit/hyperactivity disorder. *Journal of Developmental and Behavioral Pediatrics: JDBP, 37*(5), 395.

Coghill, D. R., Seth, S., Pedroso, S., Usala, T., Currie, J., & Gagliano, A. (2014). Effects of methylphenidate on cognitive functions in children and adolescents with attention-deficit/hyperactivity disorder: Evidence from a systematic review and a meta-analysis. *Biological Psychiatry, 76*(8), 603–615. https://doi.org/10.1016/j.biopsych.2013.10.005

d'Angelo, L. C., Savulich, G., & Sahakian, B. J. (2017). Lifestyle use of drugs by healthy people for enhancing cognition, creativity, motivation and pleasure. *British Journal of Pharmacology, 174*(19), 3257–3267. https://doi.org/10.1111/bph.13813

Darracq, L., Blanc, G., Glowinski, J., & Tassin, J.-P. (1998). Importance of the noradrenaline–dopamine coupling in the locomotor activating effects of D-amphetamine. *Journal of Neuroscience, 18*(7), 2729–2739.

Dietz, P., Striegel, H., Franke, A. G., Lieb, K., Simon, P., & Ulrich, R. (2013). Randomized response estimates for the 12-month prevalence of cognitive-enhancing drug use in university students. *Pharmacotherapy, 33*(1), 44–50. https://doi.org/10.1002/phar.1166

Elsey, J. W. B. (2017). Psychedelic drug use in healthy individuals: A review of benefits, costs, and implications for drug policy. *Drug Science, Policy and Law, 3*, 2050324517723232. https://doi.org/10.1177/2050324517723232

Emanuel, R. M., Frellsen, S. L., Kashima, K. J., Sanguino, S. M., Sierles, F. S., & Lazarus, C. J. (2013). Cognitive enhancement drug use among future physicians: Findings from a multi-institutional census of medical students. *Journal of General Internal Medicine, 28*(8), 1028–1034.

Farah, M. J., Haimm, C., Sankoorikal, G., Smith, M. E., & Chatterjee, A. (2009). When we enhance cognition with Adderall, do we sacrifice creativity? A preliminary study. *Psychopharmacology, 202*(1–3), 541–547. https://doi.org/10.1007/s00213-008-1369-3

Franke, A. G., Gränsmark, P., Agricola, A., Schühle, K., Rommel, T., Sebastian, A., … Lieb, K. (2017). Methylphenidate, modafinil, and caffeine for cognitive enhancement in chess: A double-blind, randomised controlled trial. *European Neuropsychopharmacology, 27*(3), 248–260.

Franke, A. G., Roser, P., Lieb, K., Vollmann, J., & Schildmann, J. (2016). Cannabis for cognitive enhancement as a new coping strategy? Results from a survey of students at four universities in Germany. *Substance Use & Misuse, 51*(14), 1856–1862. https://doi.org/10.1080/10826084.2016.1200619

Fredholm, B. B., Yang, J., & Wang, Y. (2017). Low, but not high, dose caffeine is a readily available probe for adenosine actions. *Molecular Aspects of Medicine, 55*, 20–25. https://doi.org/10.1016/j.mam.2016.11.011

Holloway, K., & Bennett, T. (2012). Prescription drug misuse among university staff and students: A survey of motives, nature and extent. *Drugs-Education Prevention and Policy, 19*(2), 137–144. https://doi.org/10.3109/09687637.2011.594114

Ilieva, I., Boland, J., & Farah, M. J. (2013). Objective and subjective cognitive enhancing effects of mixed amphetamine salts in healthy people. *Neuropharmacology, 64*, 496–505. https://doi.org/10.1016/j.neuropharm.2012.07.021

Jeffers, A. J., & Benotsch, E. G. (2016). Non-medical use of ADHD stimulants for appetite suppression and weight loss. In *Chemically Modified Bodies* (pp. 149–172). Springer.

Linssen, A. M. W., Sambeth, A., Vuurman, E. F. P. M., & Riedel, W. J. (2014). Cognitive effects of methylphenidate in healthy volunteers: A review of single dose studies. *International Journal of Neuropsychopharmacology, 17*(6), 961–977. https://doi.org/10.1017/S1461145713001594

Maier, L. J., Ferris, J. A., & Winstock, A. R. (2018). Pharmacological cognitive enhancement among non-ADHD individuals—A cross-sectional study in 15 countries. *International Journal of Drug Policy, 58*, 104–112.

Maier, L. J., Haug, S., & Schaub, M. P. (2016). Prevalence of and motives for pharmacological neuroenhancement in Switzerland—Results from a national internet panel. *Addiction, 111*(2), 280–295.

Maier, L. J., & Schaub, M. P. (2015). The use of prescription drugs and drugs of abuse for neuroenhancement in Europe. *European Psychologist, 20*, 155–166.

Maier, L. J., Liakoni, E., Schildmann, J., Schaub, M. P., & Liechti, M. E. (2015). Swiss university students' attitudes toward pharmacological cognitive enhancement. *PLoS ONE, 10*(12), e0144402.

Maier, L. J., Liechti, M. E., Herzig, F., & Schaub, M. P. (2013). To dope or not to dope: Neuroenhancement with prescription drugs and drugs of abuse among Swiss university students. *PLoS ONE, 8*(11), e77967.

Mazanov, J., Dunn, M., Connor, J., & Fielding, M. L. (2013). Substance use to enhance academic performance among Australian university students. *Performance Enhancement & Health, 2*(3), 110–118.

McCabe, S. E., Knight, J. R., Teter, C. J., & Wechsler, H. (2005). Non-medical use of prescription stimulants among US college students: Prevalence and correlates from a national survey. *Addiction, 100*(1), 96–106.

Moeller, S. J., Honorio, J., Tomasi, D., Parvaz, M. A., Woicik, P. A., Volkow, N. D., & Goldstein, R. Z. (2014). Methylphenidate enhances executive function and optimizes prefrontal function in both health and cocaine addiction. *Cerebral Cortex (New York, NY), 24*(3), 643–653. https://doi.org/10.1093/cercor/bhs345

Parkes, J. D., & Fenton, G. W. (1973). Levo (–) amphetamine and dextro (+) amphetamine in the treatment of narcolepsy. *Journal of Neurology, Neurosurgery & Psychiatry, 36*(6), 1076–1081.

Pelham, W. E., Aronoff, H. R., Midlam, J. K., Shapiro, C. J., Gnagy, E. M., Chronis, A. M., … Waxmonsky, J. (1999). A comparison of ritalin and adderall: Efficacy and time-course in children with attention-deficit/hyperactivity disorder. *Pediatrics, 103*(4), e43.

Rai, D., Bhatia, G., Sen, T., & Palit, G. J. J. (2003). Anti-stress effects of Ginkgo biloba and Panax ginseng: A comparative study. *Journal of Pharmacological Sciences, 93*(4), 458–464.

Ram, S., Hussainy, S., Henning, M., Jensen, M., & Russell, B. (2016). Prevalence of cognitive enhancer use among New Zealand tertiary students. *Drug and Alcohol Review, 35*(3), 345–351.

Robertson, P., Jr., & Hellriegel, E. T. (2003). Clinical pharmacokinetic profile of modafinil. *Clinical Pharmacokinetics, 42*(2), 123–137. https://doi.org/10.2165/00003088-200342020-00002

Rosen, W., & Weil, A. (2004). *From chocolate to morphine: Everything you need to know about mind-altering drugs*. Boston, MA: Houghton Mifflin.

Rush, C. R., Kelly, T. H., Hays, L. R., Baker, R. W., & Wooten, A. F. (2002). Acute behavioral and physiological effects of modafinil in drug abusers. *Behavioural Pharmacology, 13*(2), 105–115.

Sahakian, B. J., Bruhl, A. B., Cook, J., Killikelly, C., Savulich, G., Piercy, T., … Jones, P. B. (2015). The impact of neuroscience on society: Cognitive enhancement in neuropsychiatric disorders and in healthy people. *Philosophical Transactions of the Royal Society of London. Series B, Biological Sciences, 370*(1677), 20140214. https://doi.org/10.1098/rstb.2014.0214

Sahakian, B. J., & Morein-Zamir, S. (2011). Neuroethical issues in cognitive enhancement. *Journal of Psychopharmacology, 25*(2), 197–204.

Schelle, K. J., Olthof, B. M., Reintjes, W., Bundt, C., Gusman-Vermeer, J., & van Mil, A. C. (2015). A survey of substance use for cognitive enhancement by university students in the Netherlands. *Frontiers in Systems Neuroscience, 9*, 10. https://doi.org/10.3389/fnsys.2015.00010

Singh, I., Bard, I., & Jackson, J. (2014). Robust resilience and substantial interest: A survey of pharmacological cognitive enhancement among university students in the UK and Ireland. *PLoS ONE, 9*(10), e105969. https://doi.org/10.1371/journal.pone.0105969

Stoops, W. W., Lile, J. A., Fillmore, M. T., Glaser, P. E., & Rush, C. R. (2005). Reinforcing effects of modafinil: Influence of dose and behavioral demands following drug administration. *Psychopharmacology, 182*(1), 186–193. https://doi.org/10.1007/s00213-005-0044-1

Storebø, O. J., Krogh, H. B., Ramstad, E., Moreira-Maia, C. R., Holmskov, M., Skoog, M., … Gluud, C. (2015). Methylphenidate for attention-deficit/hyperactivity disorder in children and adolescents: Cochrane systematic review with meta-analyses and trial sequential analyses of randomised clinical trials. *BMJ, 351*. https://doi.org/10.1136/bmj.h5203

Tsai, H. H., Lin, H. W., Simon Pickard, A., Tsai, H. Y., & Mahady, G. J. I. (2012). Evaluation of documented drug interactions and contraindications associated with herbs and dietary supplements: A systematic literature review. *International Journal of Clinical Practice, 66*(11), 1056–1078.

Turner, D. C., Robbins, T. W., Clark, L., Aron, A. R., Dowson, J., & Sahakian, B. J. (2003). Cognitive enhancing effects of modafinil in healthy volunteers. *Psychopharmacology, 165*(3), 260–269.

Vargo, E. J., & Petróczi, A. (2016). "It Was Me on a Good Day": Exploring the Smart Drug Use Phenomenon in England. *Frontiers in Psychology, 7*, 779. https://doi.org/10.3389/fpsyg.2016.00779

Vrecko, S. (2013). Just how cognitive is "cognitive enhancement"? On the significance of emotions in university students' experiences with study drugs. *AJOB Neuroscience, 4*(1), 4–12. https://doi.org/10.1080/21507740.2012.740141

Ward, W. E., Chilibeck, P. D., Comelli, E. M., Duncan, A. M., Phillips, S. M., Robinson, L. E., & Stellingwerff, T. (2019). Research in nutritional supplements and nutraceuticals for health, physical activity, and performance: Moving forward. *Applied Physiology, Nutrition, and Metabolism, 44*(5), 455–460.

White, B. P., Becker-Blease, K. A., & Grace-Bishop, K. (2006). Stimulant medication use, misuse, and abuse in an undergraduate and graduate student sample. *Journal of American College Health, 54*(5), 261–268. https://doi.org/10.3200/JACH.54.5.261-268

Wolff, W., Brand, R., Baumgarten, F., Lösel, J., & Ziegler, M. (2017). Modeling students' instrumental (mis-) use of substances to enhance cognitive performance. *BioPsychoSocial Medicine, 8*, 1–11.

Zhang, C., Jiang, H., Liu, F., & He, Y. (2017). Application of near-infrared hyperspectral imaging with variable selection methods to determine and visualize caffeine content of coffee beans. *Food and Bioprocess Technology, 10*(1), 213–221.

2

Psychosocial Motivators of "Smart Drug" Use Amongst University Students

Robert C. Dempsey

The Prevalence of "Smart Drug" Use Amongst University Students

The overall rate of prescription of cognitive enhancing "smart drug" substances like Ritalin and modafinil (Cakic, 2009; Giurgea & Salama, 1977) appears to have increased over the past ten or so years (e.g. Piper et al., 2018; Renoux, Shin, Dell'Aniello, Fergusson, & Suissa, 2016). Establishing the actual prevalence of non-prescribed stimulant use amongst students is, however, a difficult endeavour. There is a limited understanding of the prevalence of university students' use of non-prescribed stimulants, especially outside of the USA (Ragan, Bard, & Singh, 2013), although students appear to be more likely to use non-prescribed stimulants compared to their same-age peers not in university/college (Ford & Pomykacz, 2016). There is also some empirical evidence to suggest that use of Ritalin

R. C. Dempsey (✉)
Manchester Metropolitan University, Manchester, UK
e-mail: r.dempsey@mmu.ac.uk

© The Author(s) 2021
M. Hall et al. (eds.), *Chemically Modified Minds*,
https://doi.org/10.1007/978-981-15-6771-1_2

and similar substances to improve academic performance has not increased since the 1960s despite media reports to the contrary (Rosiers & Van Hal, 2010). It is possible that "smart drug" use amongst university students is less widespread than commonly assumed.

Many students who use smart drugs appear to do so on an intermittent basis (McCabe, Teter, & Boyd, 2006) and tend to secure these substances via classmates and friends, who may themselves have legitimate prescriptions and medical reasons for their use (Bavarian et al., 2017; Garnier-Dykstra, Caldeira, Vincent, O'Grady, & Arria, 2012; McCabe et al., 2006; Vrecko, 2015). Research has suggested that around half of students who have been legitimately prescribed stimulant medications (e.g. for a diagnosis of attention-deficit hyperactivity disorder) have been approached by other students to sell, trade, or pass on their medication (McCabe et al., 2006). Longitudinal data has also suggested that around 60% of university students have been offered a prescription stimulant during the course of their studies (Garnier-Dykstra et al., 2012), indicating a potential issue with the availability of such stimulants to students. Alternative sources of smart drugs can include family members, clinicians deceived into prescribing stimulants, to more illicit sources such as university "black markets" where students sell cognitive enhancers to other students (Vrecko, 2015), online via the Dark Web (Cunliffe, Décary-Hêtu, & Pollak, 2019; Del Vigna et al., 2016), and unregulated internet pharmacies (Ragan et al., 2013). Given that obtaining smart drugs may involve illegal behaviour, which users may be unwilling to openly declare, ascertaining the true rates of student smart drug use may be difficult to determine.

Estimates of the prevalence of lifetime, recent, and active smart drug use amongst students have been provided by a number of published empirical studies. These studies typically suggest rates of any lifetime use of non-prescribed cognitive enhancing medications for academic purposes as being around 6–7%, with rates varying across countries (e.g. Helmer et al., 2016; Lucke et al., 2018; McCabe, 2008; McCabe, Knight, Teter, & Wechsler, 2005). Rates of recent use over the past 12 months are lower compared to lifetime use at around 4% (Lucke et al., 2018; McCabe et al., 2005), and around 2% in the past month amongst US students (McCabe et al., 2005). Most research has tended to focus on undergraduate students, although there is some evidence to suggest similar rates of lifetime use exist amongst graduate students (Verdi, Weyandt, & Zavras, 2016).

There are, however, some well-known issues with understanding the prevalence of non-prescribed stimulants amongst students. There are inconsistencies across studies in how use is assessed and defined, and whether stimulant use includes recreational use in addition to intended use for promoting cognitive and academic performance (Ford & Pomykacz, 2016; Hall et al., 2005; Ragan et al., 2013; Schleim, 2010). Notably higher lifetime rates of non-prescribed stimulant use have been reported in student samples when the reasons for use are unspecified or include both recreational and academic use (Carter, Peralta, & Xi, 2019; Silvestri & Correia, 2016), for example "for non-medical purposes" (Bavarian et al., 2017) or "illegal use" (DeSantis, Webb, & Noar, 2008). Reported lifetime usage rates reported in such studies have included 8% (McCabe et al., 2006; Teter, Esteban, Cranford, Boyd, & Guthrie, 2005), 13.7% (Hall et al., 2005), 17% (Bavarian et al., 2017), 31% (Garnier-Dykstra et al., 2012), and 34% (DeSantis et al., 2008). Clearly, there is a need for empirical studies and surveys to focus on students' intended non-prescribed stimulant use for academic reasons and avoid the possible inflation of prevalence rates by conflating these statistics with usage for more recreational purposes.

Whilst the use of non-prescribed "smart drugs" may be relatively low in the student population (i.e. is not a majority behaviour), awareness that these substances may be used to promote cognitive function and academic performance is likely to be significantly higher (Weyandt et al., 2009). For example, Maier, Liechti, Herzig, and Schaub's (2013) study suggested that a significant majority (93.7%) of their Swiss university student sample were aware that prescription stimulants could be used to improve their cognitive function. Sixty per cent of Weyandt et al.'s (2009) sample reported knowing of other students who misuse non-prescription stimulants. The high awareness of these substances amongst students remains a concern for preventing use.

Generally, students report use of non-prescribed stimulants to improve their cognitive and academic functioning (e.g. Bavarian et al., 2017; Kerley, Copes, & Griffin, 2015; Verdi et al., 2016). It should be noted, however, that a variety of substances aside from non-prescribed stimulants may be used by students to aid their academic performance, including various legal and illegal substances, over-the-counter medicines, and medications which impact other aspects of physiological functioning. Common legal substances which students use to improve their academic performance include coffee, tea, caffeine pills, energy drinks, Omega 3 supplements, over-the-counter cold and flu tablets, alcohol, and tobacco (Lucke et al., 2018; Maier et al., 2013; Maier & Schaub, 2015). The reported rates of the use of such legal substances to promote academic performance remain higher compared to those reported for non-prescribed stimulants (e.g. 46.6; Lucke et al., 2018). Common illicit substances used to aid performance typically include cannabis, followed by cocaine, other forms of amphetamines, and speed and crystal methamphetamine (Lucke et al., 2018). There is also evidence that students may use non-prescribed sedatives and sleeping medications to improve next-day cognitive functioning as an aid to relaxation, known as "indirect cognitive enhancement" (Lehne et al., 2018; Maier et al., 2013), potentially in combination with non-prescribed stimulants.

Whilst university students may use a variety of non-prescribed stimulants to aid their academic performance, these substances are also accompanied with a number of side effects which may hinder their learning (Maier & Schaub, 2015). For example, a large-scale survey study of Swiss university students' use of prescription medicines and drugs of abuse for cognitive enhancement reported that common negative consequences of use included nervousness, disordered sleep, headaches, depressive symptoms, loss of appetite, tachycardia, anxiety attacks, and aggressive behaviour, with 5% of the sample also reporting problems with their education as a result of their stimulant use (Maier et al., 2013). Although, a significant proportion of the sample in Maier et al.'s (2013) study (38.1%) reported experiencing no problems related to using neuroenhancing substances. There are also potential risks with sourcing "smart drugs" online, including whether the substances obtained are counterfeits and/or whether these substances contain innocuous or harmful compounds due to their illicit, unregulated nature. It is not unreasonable to expect some potentially serious reactions to counterfeit cognitive enhancing stimulants obtained online.

In sum, whilst the use of "smart drugs" by university students appears to be a minority behaviour, there is evidence to suggest that a clear majority of the student population are aware that these non-prescribed substances can be taken as a potential means of improving their academic performance. Despite this, there are several potentially severe side effects which may accompany the use of these non-prescribed stimulant medications, including negative effects on students' academic achievement in addition to various health-related consequences, indicating that using these substances is not without risk.

Why Do University Students Use "Smart Drugs"? A Review of Psychosocial Motivators for Students' Smart Drug Use

Improving Academic Performance

Perhaps unsurprisingly, university students appear to be largely motivated to use smart drugs to improve their general academic performance and learning (Kerley et al., 2015; Maier et al., 2013; Weyandt et al., 2013). Related motivations include improving students' focus during teaching sessions, improving test performance and coursework grades (e.g. London-Nadeau, Chan, & Wood, 2019; Verdi et al., 2016; Weyandt et al., 2009, 2013). Those students who perceive greater academic benefits associated with the use of non-prescribed stimulants for academic performance have been reported to have higher odds of personally using such smart drugs themselves (Arria et al., 2018). More specifically, students tend to endorse using smart drugs to improve their cognitive abilities in the context of their studies, such as improving their memory recall, alertness, and concentration span (Bavarian et al., 2017; DeSantis, Noar, & Webb, 2010; Teter et al., 2005). Other research has suggested that motivations for using prescription stimulants change over the course of university studies, such as moving from initial curiosity towards more academic performance-related motives (Garnier Dykstra et al., 2012).

In contrast to many students' expectations regarding the potential benefit of smart drug use for academic performance, longitudinal evidence suggests no clear benefit in terms of academic outcomes (namely grade point averages) associated with non-prescribed stimulant use (Arria et al., 2017). Rather, academic performance was significantly improved over time amongst those who abstained from using non-prescribed stimulants (Arria et al., 2017). Lower academic performances (i.e. lower grade point averages) have also been associated with greater likelihoods of non-prescribed stimulant use (McCabe et al., 2005, 2006), which could indicate a perceived pressure to use such "smart drugs" as a means to improve one's academic performance. A separate study reported that university students who used non-medically prescribed stimulants had poorer academic performance at the end of their first year of study and also studied less, missed more scheduled teaching, and socialised more than non-users (Arria, O'Grady, Caldeira, Vincent, & Wish, 2008). Although, only a minority (14%) of the student sample in Hall et al.'s (2005) study agreed that use of illicit stimulants had long-term positive effects on their academic performance. Whilst students may be motivated to use non-prescribed stimulants to improve their academic performance, there is no convincing evidence that such substances are actually associated with improved grades.

To date, the majority of the research on students' motivations for using smart drugs for academic performance has been quantitative in nature. Quantitative approaches may provide a limited understanding of students' experiences and justification for using these substances, although some qualitative studies have explored students' experiences and decisions to use smart drugs. For example, students have discussed how non-prescribed stimulants like Adderall helped them to stay awake for longer to read and write assignments, and improved their focus during lectures which in turn enhanced their retention of information (Kerley et al., 2015). Qualitative studies have also indicated that some students perceive increases to their intelligence are associated with their stimulant use (DeSantis et al., 2008, 2010; Kerley et al., 2015), although other studies have suggested the contrary position, of no perceived benefit to intelligence or cognitive abilities in the longer term associated with stimulant use (London-Nadeau et al., 2019).

Non-academic Social Reasons for Use

Aside from academic reasons, students also report using non-prescribed stimulants for social and recreational purposes, including to "get high" (Teter et al., 2005), to become more talkative and sociable (DeSantis et al., 2010), and to experiment and "party longer" (Bavarian et al., 2017). Non-prescribed stimulants are often taken with other substances for recreational purposes, most commonly alcohol (DeSantis et al., 2010). Although the extent of such motives do seem to vary in the literature, with some students viewing non-prescribed stimulants as purely a means to improve their academic performance and not for recreational use (e.g. DeSantis et al., 2008, 2010). Students using stimulants for academic purposes appear to view their use as being rational and legitimate in nature, and as a valid short-term means of achieving their career aspirations, compared to those who use the same substances for recreational purposes (Kerley et al., 2015).

Common Demographic Risk Factors for Use

A number of risk factors for the use of prescription stimulants by students have been identified. For example, having an existing legitimate prescription for stimulants has been associated with higher odds of using these substances on a non-medical basis to improve academic performance (Lucke et al., 2018). There is mixed evidence on the role of year of study on usage rates (Weyandt et al., 2013), with some reports of increased use amongst students in later years of study (e.g. McCabe et al., 2006). Other studies have failed to observe differences in use based on students' current year of study (Gallucci, Martin, Hackman, & Hutcheson, 2017; Weyandt et al., 2009), suggesting no real increase in use over the course of university studies (Garnier-Dykstra et al., 2012).

In terms of sex-differences in use, male students tend to be more likely to use non-prescription stimulants than female students (Gallucci et al., 2017; Hall et al., 2005; Lucke et al., 2018; Maier et al., 2013; Rosiers & Van Hal, 2010;Teter et al., 2005 ; Weyandt et al., 2013). There is, however, some evidence to suggest female students are more likely to use "soft enhancers" like coffee, vitamins and tonics, and energy drinks, compared to male students (Maier et al., 2013). The specific reasons for this sex-difference in the use of non-prescribed stimulants are not wholly clear, and some studies have failed to observe gender-difference in motives for using non-prescription stimulants despite differences in prevalence rates (Teter et al., 2005; Weyandt et al., 2013). Other risk factors for smart drug use include being White, and being affiliated to a student society such as a sorority or fraternity (DeSantis et al., 2008; McCabe, 2008; McCabe et al., 2005, 2006; Pino, Tajalli, Smith, & DeSoto, 2017; Weyandt et al., 2009, 2013).

There is some evidence to suggest that students' place of residence influences the use of "smart drugs", with higher use amongst those living away from their parental home (Rosiers & Van Hal, 2010), off-campus in general (Bavarian et al., 2017; McCabe et al., 2006), and those who live off-campus without family members (Pino et al., 2017). The role of Greek sorority/fraternity affiliation as a risk factor for use amongst US students may be reflective of shared residences for members of these societies and the ease of obtaining smart drugs from others (DeSantis et al., 2010). Smart drug use may be more likely when living with other students without being on the immediate university premises, where use could be caught or subject to reprimands by the university. Given that students tend to obtain prescription stimulants from other students (Garnier-Dykstra et al., 2012), whether explicitly from others or potentially by theft, it is not surprising that living with other students is one risk factor for use.

Academic Pressure and Competitiveness

A growing literature body of research has linked the use of non-prescribed stimulants specifically to perceived academic pressures and competitiveness (DeSantis et al., 2008; Maier et al., 2013), with heavier use noted at universities with more competitive entry requirements (McCabe et al., 2005). Specific subgroups of students may also be at higher risk for use, particularly those studying courses associated with high stress levels and competition. For example, medical students who had stronger perceptions that medical school is competitive, and who had higher stress levels, were more likely to use non-prescribed stimulants to improve their academic performance (De Bruyn, Wouters, Ponnet, & Van Hal, 2019). Students who reported use of non-prescribed stimulants in the previous year were more likely to engage in other forms of academic dishonesty, particularly plagiarism of other students' work (Gallucci et al., 2017), which may be broadly indicative of pressures to perform well in academic studies.

Students may experience heightened pressures to perform at key assessment and examination periods and may turn to using non-prescribed stimulants at these key times of the academic year (Kerley et al., 2015). Indeed many students appear to justify their use of non-prescription stimulants only during times of heightened stress (Kerley et al., 2015). There have, however, been some mixed findings on the role of assessment periods as a key time for smart drug use. Some studies have reported increased use of smart drugs at key assessment times and examination periods (Rosiers & Van Hal, 2010), including during final examinations (DeSantis et al., 2010; Hall et al., 2005), whilst others have suggested that students tend to use softer cognitive enhancers (e.g. coffee) nearer to examinations (Maier et al., 2013). A novel study analysing twitter posts relating to Adderall, a commonly abused cognitive enhancer, also suggested peaks of posts during December and May assessment periods (Hanson et al., 2013). Whilst a common assumption is that students may increase their use of cognitive enhancing "smart drugs" at the time of assessments, the evidence to date is somewhat more mixed although few studies have focused on use of non-prescribed stimulants over the course of an academic year.

Social Norms and Social Acceptability of "Smart Drug" Use

A body of research has investigated the potential social pressures experienced by students to use non-prescribed stimulants for academic purposes. These influences can include an explicit pressure to initiate or maintain use, or an implicit perception that using smart drugs is a commonplace and an accepted behaviour amongst students on the same course or at the same institution. Indeed, a number of studies have reported that students commonly perceive "smart drug" use to be a common and widespread behaviour at their institution (DeSantis et al., 2010; Kerley et al., 2015). In a study sampling graduate students, the most frequently reported perceived motivations for smart drugs included knowing of other students using non-prescribed stimulants during examinations and whilst studying and during final assessments (Verdi et al., 2016). Indeed, associating with other stimulant-using students has been associated with higher odds for personal use of such substances for academic purposes (Lucke et al., 2018). Students who use non-prescribed stimulants often view these substances as being different from common "street drugs", which may have negative physiological effects on the user, and viewed smart drugs as a more socially acceptable and legitimate means for achieving their goals and career aspirations (Kerley et al., 2015).

A number of studies have explicitly focused on students' social normative perceptions of non-prescribed stimulants amongst their peers. Perceived social norms are a key predictor of various health-related behaviours, and whilst there are different conceptualisations for what a social norm is, they commonly focus on perceived peer use and perceived peer approval of use (Dempsey, McAlaney, & Bewick, 2018). For example, a large study with European university students reported that the majority of students thought that the majority of their peers at their university used stimulants more frequently than themselves to improve their academic performances (Helmer et al., 2016). Other studies have suggested similar misperceptions or overestimations of the use and acceptability of non-prescription stimulants (McCabe, 2008; Silvestri & Correia, 2016). These misperceptions are similar to those noted for other substances used by students (Perkins, Meilman, Leichliter, Cashin, & Presley, 1999), including alcohol (McAlaney et al., 2015), tobacco (Pischke et al., 2015), cannabis (Dempsey et al., 2016), other illicit substances (Helmer et al., 2014), and non-prescribed sedatives (Lehne et al., 2018). Perceptions, or "misperceptions", that smart drug use is more common and more accepted by one's peers may exert social pressure on students to match what they perceive the social norm is (Festinger, 1954). For example, students who perceived that their friends and family are more approving of non-prescription stimulant use for academic purposes were more likely to use these substances (Pino et al., 2017). Similarly, students who use non-prescribed stimulants in the past year perceived that their peers had greater use and approval of smart drug use than non-users (Silvestri & Correia, 2016). However, the study by Silvestri and Correia (2016) focused on general non-prescription stimulant use, rather than use for academic purposes alone.

Overestimations of the use and acceptability of non-prescription stimulants amongst students may arise for a number of reasons, such as the false consensus effect (Ross, Greene, & House, 1977). The highlighting of what is a minority behaviour by the media and through casual conversation may make such behaviours seem to be commonplace, may ignore the actual healthy behaviours of the majority, and inflate the perceived social norms of using "smart drug" substances (Dempsey et al., 2018; Maier & Schaub, 2015; Perkins, 2003). Given that non-prescription stimulant use for academic purposes appears to be a minority behaviour, these misperceptions could make "smart drugs" appear to be more normative and acceptable than the actual reported rates.

Expectancy Effects

Similar to the inflated perceptions of the social norms of "smart drug" use, there is a growing body of research suggesting that students misperceive, possibly overestimate, how effective non-prescribed stimulants will be in terms of improving their cognitive abilities and academic performance. There is evidence of a placebo or an expectancy effect associated with the use of cognitive enhancers by students. For example, a novel experimental study found that students were no better than chance at guessing whether they had actually been prescribed an enhancer (mixed-amphetamine salts) or a placebo (Cropsey et al., 2017). Students given a cognitive enhancer improved performance on only 2 out of 31 cognitive performance tasks in this study; however, those who believed that they had received the active medication, regardless of what they actually received, had improved performance on the cognitive experimental tasks (Cropsey et al., 2017). Other experimental studies with student samples (without ADHD) suggest that the benefits associated with stimulant medication may be due to expectancy effects (Lookatch, Fivecoat, & Moore, 2017), and that any benefits are more pronounced for subjective measures (e.g. positive emotion) and/or autonomic functioning (e.g. heart rate; Weyandt et al., 2018). These studies with otherwise healthy students without a history of ADHD support other research suggesting limited cognitive benefits associated with taking cognition-enhancing stimulants amongst healthy individuals (Lookatch et al., 2017).

Motivations for Not Using "Smart Drugs"

So far, this review of the psychosocial motivators of non-prescribed stimulants for academic purposes has focused on the factors associated with an increased likelihood of use. There are, however, a small number of studies which have explicitly focused on the factors associated with abstinence from using "smart drugs". For example, students who have a more academic "ethic", that is, those who prioritise their academic studies and who study in a disciplined and intense manner, are less likely to use non-prescribed stimulants (Pino et al., 2017). Factors such as students' concern over the possible negative consequences of using non-prescribed stimulants and social disapproval from friends and family have been highlighted as other motivators for abstinence from using stimulants (Rosansky & Rosenberg, 2019).

Ethnicity has been consistently identified as a possible protective factor against using "smart drugs". Various studies have found that non-White students are less likely to use non-prescribed stimulants compared to White students (e.g. Arria et al., 2008), although some studies have failed to observe differences in stimulant use between students of different ethnicities (Carter et al., 2019). However, a stronger sense of ethnic identity (relating to a closer social identification with one's ethnicity and positive self-esteem) reduced the use of non-prescription medications amongst non-White students but not for White students (Carter et al., 2019). These studies are however limited by their dichotomisation of ethnicity into groups of White versus non-White students, limiting an understanding of the experiences of specific ethnic groups (Arria et al., 2008; Carter et al., 2019). Although, some studies have reported lower rates of use amongst Asian and African American students compared to White/Caucasian students (e.g. McCabe et al., 2005; Teter et al., 2005). It may be that having stronger identification with a relevant social group who share similar cultural values, especially one which has low approval of stimulant use, is important in determining one's use of stimulants.

Summary of Motivators of "Smart Drug" Use

There are several key motivators for students' use of non-prescribed stimulant substances. Improving one's academic performance and ability to study appear to be significant motivators for use, however a range of demographic factors (e.g. being male, living off-campus with other students, affiliation with a student society), perceived pressure to succeed academically, and the expectancy that "smart drug" use will improve performance also appear to be important in the uptake of such substances. There is also a significant role for social influence factors, such as perceived social norms and pressure from peers. Students who use non-medically prescribed prescription stimulants appear to perceive that these substances are more socially acceptable and more widely used by their peers and seem to view these substances as being more effective on their academic performance than the reality. Targeting such misperceptions and faulty beliefs about these substances' effectiveness, alongside addressing their availability, appears to be important for intervention efforts to reduce students' use of stimulant "smart drugs".

How Can "Smart Drug" Use Be Discouraged Amongst University Student Populations?

There are several potential targets for intervention to reduce students' use of non-prescribed stimulants, however, a number of the reviewed risk factors for using stimulants by students are unchangeable or difficult to change (e.g. fixed demographic variables). Therefore, intervention efforts need to be focused on those psychosocial factors which can be targeted and changed (Looby, Beyer, & Zimmerman, 2015), such as students' self-efficacy, expectancies, beliefs and perceptions of the benefits of stimulant use, and the perceived social norms of use. To date, however, there have been few published interventions focusing on psychosocial factors to reduce students' usage of non-prescribed stimulants.

A novel study tested an expectancy-challenge based intervention amongst a sample of stimulant-naïve students (i.e. those without a history of use) which presented students with research evidence challenging the perceived cognitive benefits of stimulant use (Looby, De Young, & Earleywine, 2013). Students receiving the challenge intervention had weakened positive expectancies of the cognitive enhancement benefits of stimulants post-intervention, with no difference between the intervention and a no-intervention control group at a six-month follow-up. There were, however, no group differences in the initiation of non-prescription stimulant use between groups, although more negative expectancies of stimulant use relating to arousal and anxiety feelings appeared to be protective against stimulant use (Looby et al., 2013).

One potential existing intervention method for stimulant use is the Social Norms Approach, a means of promoting positive behaviour by challenging misperceptions of the perceived acceptability and use of substances amongst a clearly defined social group (Dempsey et al., 2018; Perkins, 2003). This approach has been widely used with university student groups to challenge other substance use behaviours, particularly alcohol (Dempsey et al., 2018). Such interventions challenge overestimations of peer use and attitudes towards substances through information-based interventions, primarily using web-based personalised normative feedback to highlight discrepancies between students' own attitudes and use with the perceived and actual norms. A promising approach would be using social normative feedback to discourage initiation of stimulant use amongst students by promoting the low actual rates of use amongst the student body. Indeed, previous work with students who abstain or drink low amounts of alcohol has indicated that social norms feedback can protect against time-related increases in use (Neighbors et al., 2011). To date, no studies appear to have applied this approach to reduce or prevent use of non-prescribed stimulants amongst students.

Alongside challenging the perceived social pressure and norms of stimulant use, it would also be appropriate for interventions to focus on improving students' self-efficacy and ability to resist initiating use of stimulants. As previously discussed, university students appear to perceive that such substances are readily available on campus, can be sourced from other students, and are commonly used by their peers. Improving students' confidence in their academic abilities, and improving their "academic work ethic" (Pino et al., 2017), may also be potential targets for interventions to bolster individual students' abilities to resist initiating use. In relation to this, work by Carter et al. (2019) suggests the potential benefits of a shared social identity with a group which disapproves of stimulant use. Whilst in Carter's study this focused on ethnic identity, it may also be prudent to reinforce students' sense of shared identity with other students at their university alongside intervention messages highlighting the low actual rates of stimulant use.

Alternative suggested targets for intervention from the empirical literature include highlighting the actual lack of improvements to academic performance amongst students at-risk for using prescription stimulants for academic purposes (Arria et al., 2017). Highlighting the potential risks of non-medically prescribed stimulant use has also been suggested (Arria et al., 2008), although this could be problematic as there is evidence to suggest students misperceive the likelihood of personally experiencing negative consequences of substance use (Mallett, Lee, Neighbors, Larimer, & Turrisi, 2006). In addition to these potential psychosocial factors, addressing the actual availability of stimulants on-campus is a concern for intervention efforts, especially considering the high number of students who report that such substances are available to them (e.g. Weyandt et al., 2009). Whilst the intervention literature is somewhat limited in relation to reducing use and preventing initiation of stimulant use, there are some clear targets for interventions. Focusing on reducing the availability of these substances, students' expectancies of the benefits of these substances and the perceived norms of their use, alongside improving students' academic self-efficacy, require testing in appropriately controlled interventional studies.

Conclusions and Future Directions

Despite media reports to the contrary, the use of non-medically prescribed stimulants by students to improve their academic performance appears to be a minority behaviour on university campuses. Awareness of these stimulants and perceptions that these substances could potentially improve one's cognitive abilities and academic performance does, however, appear to be more prevalent. Unsurprisingly, students who do use these substances tend to do so for the perceived benefits on their studies and performance at university. There is a lack of evidence demonstrating that using non-prescribed stimulants actually leads to improvements in academic performance. Rather, it seems that there are expectancy or placebo-like effects associated with these substances, particularly for more subjective outcomes and mood states. In terms of the empirical literature, improvements are needed in terms of how stimulant use for academic reasons is assessed in studies in order to avoid conflating recreational use with intended use for improving academic performance.

Given that most students tend to obtain these substances from other students or family members, and that such substances are perceived to be readily available and effective, intervention efforts need to focus on boosting students' academic self-efficacy, their ability to resist initiating use, and on challenging the myths that such substances are widely used by their peers. It should be noted, however, that students may use a range of licit and illicit substances to improve their academic performance, including "soft enhancers" such as caffeine and over-the-counter medications, in addition to non-pharmacological means. To date, the interventional literature focusing on reducing and/or preventing stimulant use amongst students is limited in quantity, and there is a clear need for high-quality, controlled, interventional studies. Whilst the use of non-prescription stimulants may involve illegal behaviours, and may be associated with negative health and academic outcomes, there is a need for a societal debate about which forms of cognitive enhancement are deemed to be acceptable for use amongst students (Brühl, d'Angelo, & Sahakian, 2019).

References

Arria, A. M., Caldeira, K. M., Vincent, K. B., O'Grady, K. E., Cimini, M. D., Geisner, I. M., ... Larimer, M. E. (2017). Do college students improve their grades by using prescription stimulants nonmedically? *Addictive Behaviors, 65*, 245–249.

Arria, A. M., Geisner, I. M., Cimini, M. D., Kilmer, J. R., Caldeira, K. M., Barrall, A. L., ... Larimer, M. E. (2018). Perceived academic benefit is associated with nonmedical prescription stimulant use among college students. *Addictive Behaviors, 76*, 27–33.

Arria, A. M., O'Grady, K. E., Caldeira, K. M., Vincent, K. B., & Wish, E. D. (2008). Nonmedical use of prescription stimulants and analgesics: Associations with social and academic behaviors among college students. *Journal of Drug Issues, 38*(4), 1045–1060.

Bavarian, N., McMullen, J., Flay, B. R., Kodama, C., Martin, M., & Saltz, R. F. (2017). A mixed-methods approach examining stimulant use: Findings from a northern California university. *Journal of Primary Prevention, 38*(4), 363–383.

Brühl, A. B., d'Angelo, C., & Sahakian, B. J. (2019). Neuroethical issues in cognitive enhancement: Modafinil as the example of a workplace drug? *Brain and Neuroscience Advances, 3*, 239821281881601.

Cakic, V. (2009). Smart drugs for cognitive enhancement: Ethical and pragmatic considerations in the era of cosmetic neurology. *Journal of Medical Ethics, 35*(10), 611–615.

Carter, J., Peralta, R. L., & Xi, J. (2019). The association between ethnic identity and non-medical prescription drug use among a sample of college students: Does a sense of ethnic belonging matter? *Substance Use and Misuse, 54*(2), 203–213.

Cropsey, K. L., Schiavon, S., Hendricks, P. S., Froelich, M., Lentowicz, I., & Fargason, R. (2017). Mixed-amphetamine salts expectancies among college students: Is stimulant induced cognitive enhancement a placebo effect? *Drug and Alcohol Dependence, 178*, 302–309.

Cunliffe, J., Décary-Hêtu, D., & Pollak, T. A. (2019). Nonmedical prescription psychiatric drug use and the darknet: A cryptomarket analysis. *International Journal of Drug Policy, 73*, 263–272.

De Bruyn, S., Wouters, E., Ponnet, K., & Van Hal, G. (2019). Popping smart pills in medical school: Are competition and stress associated with the misuse of prescription stimulants among students? *Substance Use and Misuse, 54*(7), 1191–1202.

Del Vigna, F., Avvenuti, M., Macciu, C., Deluca, P., Petrocchi, M., Marchetti, A., & Tesconi, M. (2016). Spotting the diffusion of new psychoactive substances over the Internet. In H. Boström, A. Knobbe, C. Soares, & P. Papapetrou (Eds.), *Advances in intelligent data analysis XV. IDA 2016 Lecture notes in Computer Science, vol 9897*. Cham: Springer.

Dempsey, R. C., McAlaney, J., & Bewick, B. M. (2018). A critical appraisal of the social norms approach as an interventional strategy for health related behaviour and attitude change. *Frontiers in Psychology, 9*, 2180.

Dempsey, R. C., McAlaney, J., Helmer, S., Pischke, C., Akvardar, Y., Bewick, B. M., … Mikolajczyk, R. T. (2016). Normative perceptions of cannabis use amongst European University students: Associations between perceived peer use and peer attitudes with personal use and attitudes. *Journal of Studies on Alcohol and Drugs, 77*(5), 740–748.

DeSantis, A., Noar, S. M., & Webb, E. M. (2010). Speeding through the frat house: A qualitative exploration of nonmedical ADHD stimulant use in fraternities. *Journal of Drug Education, 40*(2), 157–171.

DeSantis, A. D., Webb, E. M., & Noar, S. M. (2008). Illicit use of prescription ADHD medications on a college campus: A multimethodological approach. *Journal of American College Health, 57*(3), 315–324.

Festinger, L. (1954). A theory of social comparison processes. *Human Communications, 7*, 117–140.

Ford, J. A., & Pomykacz, C. (2016). Non-medical use of prescription stimulants: A comparison of college students and their same-age peers who do not attend college. *Journal of Psychoactive Drugs, 48*(4), 253–260.

Gallucci, A. R., Martin, R. J., Hackman, C., & Hutcheson, A. (2017). Exploring the relationship between the misuse of stimulant medications and academic dishonesty among a sample of college students. *Journal of Community Health, 42*(2), 287–294.

Garnier-Dykstra, L. M., Caldeira, K. M., Vincent, K. B., O'Grady, K. E., & Arria, A. M. (2012). Nonmedical use of prescription stimulants during college: Four-year trends in exposure opportunity, use, motives, and sources. *Journal of American College Health, 60*(3), 226–234.

Giurgea, C., & Salama, M. (1977). Nootropic drugs. *Progress in Neuro-Psychopharmacology, 1*(3–4), 235–247.

Hall, K. M., Irwin, M. M., Bowman, K. A., Frankenberger, W., Hall, K. M., & Jewett, D. C. (2005). Illicit use of prescribed stimulant medication among college students. *Journal of American College Health, 53*(4), 167–174.

Hanson, C. L., Burton, S. H., Giraud-carrier, C., West, J. H., Michael, D., & Hanson, B. (2013). Tweaking and tweeting: Exploring Twitter for nonmedical use of a psychostimulant drug (Adderall) Among college students. *Journal of Medical Internet Research, 15*(4), e62.

Helmer, S. M., Mikolajczyk, R. T., McAlaney, J., Vriesacker, B., van Hal, G., Akvardar, Y., … Zeeb, H. (2014). Illicit substance use among university students from seven European countries: A comparison of personal and perceived peer use and personal and perceived peer attitudes towards illicit substance use. *Preventive Medicine, 67*, 204–209.

Helmer, S. M., Pischke, C. R., Van Hal, G., Vriesacker, B., Dempsey, R. C., Akvardar, Y., … Zeeb, H. (2016). Personal and perceived peer use and attitudes towards the use of nonmedical prescription stimulants to improve academic performance among university students in seven European countries. *Drug and Alcohol Dependence, 168*, 128–134.

Kerley, K. R., Copes, H., & Griffin, O. H. (2015). Middle-class motives for non-medical prescription stimulant use among college students. *Deviant Behavior, 36*(7), 589–603.

Lehne, G., Zeeb, H., Pischke, C. R., Mikolajczyk, R., Bewick, B. M., McAlaney, J., … Helmer, S. M. (2018). Personal and perceived peer use and attitudes towards use of non-prescribed prescription sedatives and sleeping pills among university students in seven European countries. *Addictive Behaviors, 87*, 17–23.

London-Nadeau, K., Chan, P., & Wood, S. (2019). Building conceptions of cognitive enhancement: University students' views on the effects of pharmacological cognitive enhancers. *Substance Use and Misuse, 54*(6), 908–920.

Looby, A., Beyer, D. L., & Zimmerman, L. (2015). Non-medical prescription stimulant use: Investigating modifiable risk factors. *Addiction Research & Theory, 23*(2), 143–147.

Looby, A., De Young, K. P., & Earleywine, M. (2013). Challenging expectancies to prevent nonmedical prescription stimulant use: A randomized, controlled trial. *Drug and Alcohol Dependence, 132*(1–2), 362–368.

Lookatch, S. J., Fivecoat, H. C., & Moore, T. M. (2017). Neuropsychological effects of placebo stimulants in college students. *Journal of Psychoactive Drugs, 49*(5), 398–407.

Lucke, J., Jensen, C., Dunn, M., Chan, G., Forlini, C., Kaye, S., … Hall, W. (2018). Non-medical prescription stimulant use to improve academic performance among Australian university students: Prevalence and correlates of use. *BMC Public Health, 18*(1), 1270.

Maier, L. J., Liechti, M. E., Herzig, F., & Schaub, M. P. (2013). To dope or not to dope: Neuroenhancement with prescription drugs and drugs of abuse among Swiss university students. *PLoS ONE, 8*(11), e77967.

Maier, L. J., & Schaub, M. P. (2015). The use of prescription drugs and drugs of abuse for neuroenhancement in Europe: Not widespread but a reality. *European Psychologist, 20*(3), 155–166.

Mallett, K. A., Lee, C. M., Neighbors, C., Larimer, M. E., & Turrisi, R. (2006). Do we learn from our mistakes? An examination of the impact of negative alcohol-related consequences on college students' drinking patterns and perceptions. *Journal of Studies on Alcohol, 67*(2), 269–276.

McAlaney, J., Helmer, S., Stock, C., Vriesacker, B., van Hal, G., Dempsey, R. C., … Mikolajczyk, R. (2015). Personal and perceived peer use of and attitudes toward alcohol among university and college students in seven EU countries: Project SNIPE. *Journal of Studies on Alcohol and Drugs, 76*(3), 430–438.

McCabe, S. E. (2008). Misperceptions of non-medical prescription drug use: A web survey of college students. *Addictive Behaviors, 33*(5), 713–724.

McCabe, S. E., Knight, J. R., Teter, C. J., & Wechsler, H. (2005). Non-medical use of prescription stimulants among US college students: Prevalence and correlates from a national survey. *Addiction, 99*, 96–106.

McCabe, S. E., Teter, C. J., & Boyd, C. J. (2006). Medical use, illicit use and diversion of prescription stimulant medication. *Journal of Psychoactive Drugs, 38*(1), 43–56.

Neighbors, C., Jensen, M., Tidwell, J., Walter, T., Fossos, N., & Lewis, M. A. (2011). Social-norms interventions for light and nondrinking students. *Group Processes & Intergroup Relations, 14*(5), 651–669.

Perkins, H. W. (2003). The emergence and evolution of the social norms approach to substance abuse prevention. In H. W. Perkins (Ed.), *The social norms approach to preventing school and college age substance abuse: A handbook for educators, counselors, and clinicians* (pp. 3–18). San Francisco: Jossey-Bass.

Perkins, H. W., Meilman, P. W., Leichliter, J. S., Cashin, J. R., & Presley, C. A. (1999). Misperceptions of the norms for frequency of alcohol and other drug use on campus. *Journal of American College Health, 47*, 253–258.

Pino, N. W., Tajalli, H., Smith, C. L., & DeSoto, W. (2017). Nonmedical prescription drug use by college students for recreational and instrumental purposes: Assessing the differences. *Journal of Drug Issues, 47*(4), 606–621.

Piper, B. J., Ogden, C. L., Simoyan, O. M., Chung, D. Y., Caggiano, J. F., Nichols, S. D., & McCall, K. L. (2018). Trends in use of prescription stimulants in the United States and territories, 2006 to 2016. *PLoS ONE, 13*(11), e0206100.

Pischke, C. R., Helmer, S. M., McAlaney, J., Bewick, B. M., Vriesacker, B., Van Hal, G., ... Zeeb, H. (2015). Normative misperceptions of tobacco use among university students in seven European countries: Baseline findings of the 'Social norms intervention for the prevention of polydrug use' study. *Addictive Behaviors, 51*, 158–164.

Ragan, C. I., Bard, I., & Singh, I. (2013). What should we do about student use of cognitive enhancers? An analysis of current evidence. *Neuropharmacology, 64*, 588–595.

Renoux, C., Shin, J. Y., Dell'Aniello, S., Fergusson, E., & Suissa, S. (2016). Prescribing trends of attention-deficit hyperactivity disorder (ADHD) medications in UK primary care, 1995–2015. *British Journal of Clinical Pharmacology, 82*, 858–868.

Rosansky, J. A., & Rosenberg, H. (2019). University students' self-reported reasons for abstinence from prescription and non-prescription stimulants and depressants. *Drug and Alcohol Dependence, 194*, 476–481.

Rosiers, J., & Van Hal, G. (2010). Stimulant medication use among Flemish students: Results from an exploring secondary data analysis 1965–2005. *Archives of Public Health, 67*(4), 169–178.

Ross, L., Greene, D., & House, P. (1977). The "false consensus effect" an egocentric bias in social perception and attribution processes. *Journal of Experimental and Social Psychology, 13*, 279–301.

Schleim, S. (2010). Second thoughts on the prevalence of enhancement. *Biosocieties, 5*(4), 484–485.

Silvestri, M. M., & Correia, C. J. (2016). Normative influences on the nonmedical use of prescription stimulants among college students. *Psychology of Addictive Behaviors, 30*(4), 516–521.

Teter, C. J., Esteban, S., Cranford, J. A., Boyd, C. J., & Guthrie, S. K. (2005). Prevalence and motives for illicit use of prescription stimulants in an undergraduate student sample. *Journal of American College Health, 53*(6), 253–262.

Verdi, G., Weyandt, L. L., & Zavras, B. M. (2016). Non-medical prescription stimulant use in graduate students: Relationship with academic self-efficacy and psychological variables. *Journal of Attention Disorders, 20*(9), 741–753.

Vrecko, S. (2015). Everyday drug diversions: A qualitative study of the illicit exchange and non-medical use of prescription stimulants on a university campus. *Social Science and Medicine, 131*, 297–304

Weyandt, L., White, T., Gudmundsdottir, B., Nitenson, A., Rathkey, E., De Leon, K., & Bjorn, S. (2018). Neurocognitive, autonomic, and mood effects of Adderall: A pilot study of healthy college students. *Pharmacy, 6*(3), 58.

Weyandt, L. L., Janusis, G., Wilson, K. G., Verdi, G., Paquin, G., Lopes, J., … Dussault, C. (2009). Nonmedical prescription stimulant use among a sample of college students. *Journal of Attention Disorders, 13*(3), 284–296.

Weyandt, L. L., Marraccini, M. E., Gudmundsdottir, B. G., Martinez Zavras, B., Turcotte, K. D., Munro, B. A., & Amoroso, A. J. (2013). Misuse of prescription stimulants among college students: A review of the literature and implications for morphological and cognitive effects on brain functioning. *Experimental and Clinical Psychopharmacology, 21*(5), 385–407.

3

Nutraceuticals as Cognitive Enhancers

Sarah Benson and Andrew Scholey

Introduction

We live in a competitive world where people often seek to gain advantages over others. Such advantages may apply to cognitive capabilities, for instance, improving memory efficiency; or physical capabilities, such as resistance to disease. Over recent decades, in particular, capacity for physical improvement has substantially advanced, largely due to medical and technological improvements. As such, global life expectancy at birth increased by 13.5 years for males and 14.8 years for females from 1970 to 2016 (Wang et al., 2017). One of the primary factors driving this increase is the reduction in death rates among the elderly (Wilmoth, 2000). This has resulted in rising prevalence rates of age-related cognitive complaints and dementia, with 47 million dementia patients worldwide in 2015 and an expected 66 million by 2030 (Prince et al., 2013).

Despite relatively successful efforts to improve physical capabilities, maintaining or improving aspects of health related to the brain has proven to be a far greater challenge. While individuals who experience cognitive

S. Benson (✉) • A. Scholey
Swinburne University, Melbourne, VIC, Australia
e-mail: sarahbenson@swin.edu.au, ascholey@swin.edu.au

© The Author(s) 2021
M. Hall et al. (eds.), *Chemically Modified Minds*,
https://doi.org/10.1007/978-981-15-6771-1_3

complaints may seek to maintain their brain health, many people who are free from any cognitive-related problems seek to improve their cognitive functioning to meet lifestyle and career demands. Cognitive function is a term used to describe a range of neurocognitive abilities, such as memory and attention. Cognitive functions allow us to perceive, evaluate, store, manipulate and use information from our external environment and internal sources (i.e., experience, memory, concepts, thoughts, etc.), and to respond to this information.

There are numerous taxonomies of cognitive function. Some, such as the Cattell-Horn-Carroll model, arrange cognitive function into strata, ranging from broad to narrow abilities (Schneider & McGrew, 2018). The latter may include 30 or more individual tasks. More widely used models cluster cognitive functions into several main cognitive domains, for example attention, memory, executive function, language, perception and psychomotor function. Each of these domains is complex entities, for example, the memory domain includes the components of encoding, storage and retrieval and can be clustered into short-term, long-term and working memory.

The ability to perform in each cognitive domain can be measured. The development of computerised cognitive testing greatly increased sensitivity of tests in the measurement of cognitive change, such as that induced by nutraceuticals, and enabled greater accuracy in measuring cognitive performance. However, as a consensus on a single model of the components of cognition has not yet been reached, different test batteries draw on different models of cognition and target various cognitive elements.

Cognitive functioning may be enhanced by the use of cognitive enhancers, that is, anything that improves performance on one or more of the cognitive domains. Cognitive-enhancing substances may be either pharmaceutical or nutraceutical. Pharmaceutical cognitive enhancers are developed to treat medical disorders. For example, modafinil (Provigil) is a medication used to treat excessive daytime sleepiness associated with narcolepsy or shift work (Repantis, Schlattmann, Laisney, & Heuser, 2010), and is also used recreationally for cognitive-related benefits. Nutraceuticals are products derived from food sources, typically plants, which have health benefits beyond their basic nutritional value. Interestingly, the reported magnitude of positive cognitive effects of modafinil is comparable to those of the nutraceuticals, ginseng and bacopa (Neale, Camfield, Reay, Stough, & Scholey, 2013).

The global nutraceutical industry was worth an estimated $230.9 billion in 2018 and is expected to grow to reach $336.1 billion in 2023 (BCC Research, 2018). Approximately 50% of adults in the United States regularly use a nutraceutical product (Dickinson, Blatman, El-Dash, & Franco, 2014), frequently used for their purported overall enhancement to psychological health or well-being (Bailey, Gahche, Miller, Thomas, & Dwyer, 2013; Dickinson et al., 2014; Marinac et al., 2007). However, although use of nutraceuticals has a long history, only relatively recently has scientific research rigorously assessed the potential efficacy of certain nutraceuticals. As such, little is known about their efficacy and empirical evidence in support of desired outcomes is often lacking for many nutraceutical products.

Nevertheless, there is growing evidence that certain nutraceutical supplements may have cognitive-enhancing properties, in particular, ginseng, salvia and cocoa flavanols. This chapter describes the active constituents and central nervous system (CNS)-related mechanisms of action of these products. In addition, this chapter summarises the evidence pertaining to the efficacy of these products to acutely enhance cognitive functioning, drawing upon results from controlled human clinical trials.

Ginseng

Ginseng refers to the dried root of several species of the plant genus *Panax*, of the Araliaceae plant family. Extracts of ginseng have a long history of use in traditional Chinese medicine, primarily for the purposes of providing energy and aiding recovery in the ill and elderly. More recently, ginseng is used to promote vitality and prolong life, as well as to treat an array of health conditions including depression, fatigue, ageing, inflammation, internal degeneration, nausea, tumours, pulmonary problems, dyspepsia, vomiting, nervousness, stress and ulcers (Ernst, 2005; Helms, 2004; Jeong, 2002; Kiefer & Pantuso, 2003).

The most widely used species of ginseng, *Panax ginseng* (sometimes referred to as Asian ginseng), was first cultivated in Korea around 11 BC (Yun, 2001). Other types of widely used ginseng include *Panax quinquefolius* (American ginseng), *Panax notoginseng* and *Panax japonicas.*

Active Constituents

The constituents of ginseng reported as being primarily responsible for its bioactivity are its 40 or more ginsenosides. Ginsenosides are unique to ginseng species and can be isolated from various parts of the ginseng plant, although, typically from the roots. The ginsenoside content can vary depending on the species, season of harvest and extraction methods used (Liberti & Marderosian, 1978).

Ginsenosides can be classified into three groups on the basis of their chemical structure: the Panaxadiol group (Rb1, Rb2, Rb3, Rc, etc.), the Panaxatriol group (Re, Rf, Rg1, Rg2, Rh1) and the oleanolic acid group (e.g. Ro) (Tachikawa et al., 1999).

CNS Mechanisms of Action

Nitric Oxide (NO) Synthesis

NO is an enzyme present throughout the brain and is particularly abundant in the cerebellum. It is reported to play an imperative role in various cognitive functions including learning and memory (Garthwaite, 1993), and is also involved in neurotransmission (Garthwaite, Charles, & Chess-Williams, 1988). Increased NO synthesis has been repeatedly proposed as a potential mechanism of action of ginseng (e.g. Friedl, Moeslinger, Kopp, & Spieckermann, 2001; Scott, Colligan, Ren, & Ren, 2001). Ginseng has been found to cause neurogenic NO-mediated relaxation in electrically stimulated monkey cerebral artery (Toda, Ayajiki, Fujioka, & Okamura, 2001).

Neuroprotection

Neuroprotective properties of ginseng have been demonstrated in a series of vitro and in vivo studies. Constituents of ginseng have been found to be neuroprotective in various models of Huntington's (Seo et al., 2008) and Alzheimer's (Chen et al., 2012; Li et al., 2011) disease, as well as oxygen-glucose deprivation (Jiang, Miao, Song, & Jiang, 2011; Zhu, Tao, Lou, & Wu, 2010), neural oxidative damage (Ye et al., 2011), glutamate-induced excitotoxicity (Chen et al., 2010; Zhang et al., 2012), antinociception (Mogil, Shin, McCleskey, Kim, & Nah, 1998; Seo et al., 2008), ischemia (Chen, Zhou, Cao, & Hu, 2008; Lu et al., 2011; Ye et al., 2009, 2011) and apoptosis (Li et al., 2011).

Hypothalamic-Pituitary-Adrenal (HPA) System Regulation

Early animal research indicates that both oral administration (Fllaketov, Bogdanova, Podvigina, & Bodganov, 1988) and interperitoneal injection (Hiai, Yokoyama, Oura, & Kawashima, 1983) of ginseng can increase corticosterone and plasma levels of adrenocorticotropic (ACTH), both of which are hormones released in response to stress. Conversely, another animal model study found that ginseng injected intracerebroventricularly inhibited stress-induced plasma corticosterone levels. However, these inhibitory actions were blocked by co-administration of an inhibitor of nitric oxide synthase (Nomega-Nitro-L-arginine methyl ester) suggesting that the inhibiting effects of ginseng are driven by inducing NO production in the brain (Kim et al., 1998).

Ginseng has also been found to inhibit calcium ion (Ca^{2+}) currents and cell membrane capacitance in rat adrenal chromaffin cells following ginseng. Findings suggest that ginseng saponins regulate catecholamine secretion from the adrenal chromaffin cells, and this regulation may be the cellular basis of reduced stress following ginseng (Kim, Lee, Goo, & Nah, 1998).

Modulation of Neurotransmission

Cholinergic signalling is associated with aspects of memory, attention, motivation and mood (Lopresti, 2017). Cholinergic degeneration is observed in Alzheimer's disease (AD), leading to the possibility that pro-cholinergic agents may be effective in the treatment of AD (Kumar, Singh, & Ekavali, 2015). The breakdown of acetylcholine is primarily facilitated by the enzyme acetylcholinesterase (AChE), leading to the development of the cholinesterase inhibitor family of drugs. A number of animal studies have reported improvement to scopolamine-induced memory deficits following ginseng (Hsieh, Peng, Wu, & Wang, 2000; Jin, Park, Nam, Park, & Jung, 1999; Nitta, Matsumoto, Shimizu, Ni, & Watanabe, 1995; Sloley et al., 1999). Although the mode of action remains unknown, the most consistent evidence currently available indicates that these effects may be caused by enhancement of cholinergic neurotransmission resulting from both an increase in cholinergic receptor density and acetylcholine release (Sloley et al., 1999).

A series of studies have identified cholinergic properties associated with ginseng, including increased choline acetyltransferase levels (Salim, McEwen, & Chao, 1997; Zhang, Qu, Liu, & Deng, 1990), interactions between ginseng and nicotinic receptor subtypes (Sala et al., 2002), acetylcholine release and reuptake, and the number of choline uptake sites in the hippocampus and cortex (Benishin, 1992).

Ginseng administration has also been shown to increase central dopamine and norepinephrine, and increase serotonin levels in the cortex (Petkov, 1978), including via the modulation of activity at presynaptic and postsynaptic dopaminergic receptors (Kim, Kang, Seong, Nam, & Oh, 1995).

Human Cognitive Effects Following Acute Administration

Using randomised, double-blind, placebo-controlled and balanced cross-over designs, a series of studies have assessed the effects of acute ginseng administration on cognitive functioning in healthy humans. In the first of the studies, 20 participants were administered with 200 mg, 400 mg and 600 mg of ginseng and placebo (Kennedy, Scholey, & Wesnes, 2001a). Cognitive assessments were completed as baseline, and 1, 2.5, 4 and 6 hours after condition administration. The 200 and 600 mg doses resulted in improved reaction time on attention tasks at the latest testing time-point only. Meanwhile, the 400 mg dose improved quality of memory and measures of secondary memory.

In the second study, an identical methodology was used as described in the study above; however, doses of 320, 640 and 960 mg were administrated (Kennedy, Scholey, & Wesnes, 2001b). Memory performance was significantly improved following the 960 mg dose, with the effects isolated to secondary memory. Additionally, the 320 and 640 mg doses resulted in improved speed on attention tasks.

A later study assessed the effects of 200, 400 and 600 mg ginseng on performance on a serial subtraction mental arithmetic task (Scholey & Kennedy, 2002). Results indicated that the 400 mg dose improved accuracy on the task, while the 200 mg dose improved speed to perform the task. In contrast, a study administering 400 mg and placebo to 30 participants found that 90 minutes after ingestion, ginseng improved speed of attention but not quality of attention, speed of memory or quality of memory (Sünram-Lea, Birchall, Wesnes, & Petrini, 2005). Lastly, in a study assessing the effects of 100, 200 and 400 mg of Cereboost™ (*P. quinquefolius* standardised to 10.65% ginsenosides) in 32 participants 1, 3 and 6 hours following administration found improvements to working memory at each time point in each condition. Additionally, accuracy on a choice reaction time task was improved by the 100 mg dose.

Salvia

Salvia, commonly known as sage, has been used for cognitive enhancement dating back to the ancient Greeks. Various species of *salvia* have been traditionally used to treat a range of health issues including digestive and circulation problems, bronchitis, coughs, asthma, memory problems, angina, mouth and throat inflammation, depression and excessive sweating (Lopresti, 2017). The *Salvia* genus contains around 900 species.

Common *Salvia* species include *Salvia officinalis* (common sage), *Salvia miltiorrhiza* (Chinese sage), *Salvia lavandulaefolia* (Spanish sage), *Salvia fruticose* (Greek sage), *Salvia sclarea* (clary sage) and *Salvia hispanica* (chia). The two most commonly studied species in the context of cognitive enhancement are *Salvia officinalis* and *Salvia lavandulaefolia.*

Active Constituents

Salvia plants contain a vast range of active compounds, including phenolic acids and flavonoids. The phenolic acids include caffeic acid, rosmarinic acid, salvianolic acids, sagecoumarin, lithospermic acid, sagerinic acid and yunnaneic acid. The most prevalent flavonoids include luteolin, apigenin, hispidulin, kaempferol and quercetin (Lopresti, 2017). *Salvia* plants are also rich in essential oils, including terpenoids with α and β-thujone, camphor, 1,8-cineoles, α-humulene, β-caryophyllene and viridiflorol. They also contain diterpenes and triterpenes such as carnosic acid, ursolic acid, carnosol and tanshinones.

Constituents vary considerably across *salvia* species. For example, the level of rosmarinic acid is high in *Saliva officinalis* but low in *Salvia hypoleuka* (Shekarchi, Hajimehdipoor, Saeidnia, Gohari, & Hamedani, 2012) and thujone is high in *Saliva officinalis* but low in *Salvia lavandulaefolia.*

CNS Mechanisms of Action

Amyloid-β Peptide (Aβ)

Aβ production and deposition are widely believed to be central to the aetiology of Alzheimer's disease. It is theorised that accumulated Aβ contributes to the progressive nature of Alzheimer's disease, as the unregulated build-up of Aβ is neurotoxic and causes dysfunction to cholinergic neurons and calcium homeostasis. Aβ is known to cause learning and memory impairment and its administration in animals induces memory loss (More, Kumar, Cho, Yun, & Choi, 2016).

Salvia has been shown to protect murine Aβ-induced neurotoxicity by inhibiting increases in necrosis factor-α (TNF-α), interleukin-6 (IL-6) levels and acetylcholinesterase (AChE) activity (Teng et al., 2014). Tanshinone IIA (Tan IIA), a constituent of *Salvia*, has been found to protect against Aβ-induced neurotoxicity by inhibiting upregulation of various genes expected to be associated with Alzheimer's disease, namely, inducible nitric oxide synthase (iNOS), matrix metalloproteinase-2 (MMP-2) and nuclear transcription factor kappa (NF-κBp65, Jiang, Li, Xiang, & Jiao, 2014).

Furthermore, Aβ injected rats treated with *Salvia* have been reported *to* display behavioural improvements in learning and memory performance (Khodagholi & Ashabi, 2013). Protective effects from Aβ toxicity have also been found following administration of several isolated *Salvia* constituents, including rosmarinic acid (Alkam, Nitta, Mizoguchi, Itoh, & Nabeshima, 2007), salvianolic acid (Lee et al., 2013), carnosic acid (Rasoolijazi et al., 2013) and quercetin (Patil et al., 2003).

Cholinergic Activity

Both in vitro and animal studies show that *Salvia* is an effective AChE inhibitor (Foolad & Khodagholi, 2013; Kennedy et al., 2006; Scholey et al., 2008; Smach, Hafsa, Charfeddine, Dridi, & Limem, 2015). Protection against AChE activity has also been found following administration of several *Salvia* constituents, including rosmarinic acid, carnosic acid and quercetic (Marcelo et al., 2013; Merad et al., 2014; Sallam, Mira, Ashour, & Shimizu, 2016).

Neurotrophins

Neurotrophins are a family of proteins important to the regulation of neural survival, development, function and plasticity (Huang & Reichardt, 2001). Brain-derived neurotrophic factor (BDNF) has received particular research attention due to its role in supporting the survival of existing neurons, encouraging growth and differentiation of new neurons and synapses and enhancing learning and memory (Bowling, Bhattacharya, Klann, & Chao, 2016). Administration of *Salvia* to mice has been reported to mitigate Aβ-induced reductions in BDNF (Teng et al., 2014). Positive effects on BDNF have also been found following rosmarinic acid (Fonteles et al., 2016; Jin, Liu, Yang, Zhang, & Miao, 2013), luteolin (Xu et al., 2013) and caffeic acid (Takeda et al., 2006) administration.

The production of nerve growth factor, another neuropeptide important in the regulation of growth, maintenance, and survival of certain neurons, has been reported to be enhanced by carnosic acid and carnosol (Kosaka & Yokoi, 2003), tanshinones (Zhao et al., 2015) and quercetic (Wang et al., 2011).

Human Cognitive Effects Following Acute Administration

A series of double-blind, placebo-controlled and crossover studies have assessed the cognitive effects of acute *Salvia* administration in healthy human participants. In an assessment of 50, 100 and 150 μL *salvia* essential oil extract and placebo, 20 participants underwent cognitive testing 1, 2.5, 4 and 6 hours following administration (Tildesley et al., 2003). Performance on immediate and delayed memory tasks was improved by the 50 μL dose at 1 and 2.5 hours and the 100 μL dose at 2.5 hours. Using the same methodology, 24 participants were administered 25 and 50 μL *salvia* essential oil extract and placebo (Tildesley et al., 2003). The 50 μL dose improved immediate word recall 1 hour as well as 4 hours post-dose. A third study, again using the same methodology, assessed 25 and 50 μL *salvia* essential oil extract and placebo in 24 participants, but using a more comprehensive cognitive testing battery. Results showed that both the 25 and 50 μL doses increased the 'speed of memory' component of the cognitive testing and the 25 μL dose also resulted in improvement on the 'secondary memory' factor.

In an assessment of 300 and 600 mg dried sage leaf and placebo, 30 participants completed a multitasking cognitive battery 1 and 4 hours post-dose (Kennedy et al., 2006). Task performance was significantly improved by the 600 mg dose at both testing time points. However, the 300 mg dose reduced performance at 4 hours post-dose. In a study of 20 participants administered 167, 333, 666 and 1332 mg *Salvia* extract and placebo, cognitive assessments were completed 1, 2.5, 4 and 6 hours post-dose (Scholey et al., 2008). Results demonstrated that the 333 mg dose resulted in improved secondary memory performance at each testing time point which the same outcome was improved by a lesser extend following 167 mg at 2.5 and 4 hours post-dose and 1332 mg at 4 hours. Accuracy of attention was enhanced by the 333 mg dose at 1, 4 and 6 hours post-dose. Speed of memory, speed of attention and working memory were not affected by any of the conditions. In a study of 36 participants who were administered 50 μL *Salvia* essential oil and placebo and assessed on cognitive performance at one and four hours post-dose, improvements were found on secondary memory and attention tasks, most noticeably at the one-hour post-dose testing time (Kennedy et al., 2011).

Lastly, using a single-blinded study design, 135 participants underwent cognitive testing while in a room scented with one of two *salvia* extracts (*Salvia officinalis* and *Salvia lavandulaefolia*) or no aroma (Moss, Rouse, Wesnes, & Moss, 2010). Results showed that the *salvia* officinalis group performed better than the no aroma on quality of memory and secondary memory components.

Cocoa Flavanols

Cocoa products, especially chocolate, have received a great deal of interest regarding their effects on cognitive functioning. Reports on chocolate's health benefits date back to Aztec and Maya medical practice (e.g. Hurst, Tarka Jr, Powis, Valdez Jr, & Hester, 2002); however, scientific interest in the purported health benefits of cocoa and chocolate did not develop until the early 2000s. Most studies so far have assessed the effects of cocoa and chocolate on the cardiovascular system, which eventually resulted in endorsement from the European Food Safety Authority (EFSA, EFSA Panel on Dietetic Products NaAN, 2012) that dark chocolate, with its high flavanol content, effects the 'maintenance of normal endothelium-dependent vasodilation'.

Active Constituents

Flavonoids are natural compounds found in plants, grapes, red wine, apples and tea and are particularly abundant in cocoa (Gu et al., 2004). Cocoa contains a high concentration of polyphonic compounds and a unique combination of epicatechin, catechin and oligomeric procyanidins (Lazarus, Hammerstone, & Schmitz, 1999). Epicatechin and catechin are monomeric flavanols, while procyanidins are a class of flavonoids that are oligomeric compounds formed from catechin and epicatechin molecules. Cocoa also contains theobromine and caffeine, which have known psychoactive properties (Bell, Lamport, Butler, & Williams, 2015).

CNS Mechanisms of Action

Neuroprotection

Flavonoids have been reported to protect neurons against damage caused by neurotoxins and counteract neuronal damage underlying diseases such as Parkinson and Alzheimer diseases. Flavonoids interact with several important neuronal signalling pathways in the brain that lead to an inhibition of apoptosis (death of cells) triggered by neurotoxicity (Spencer, 2009). These include, most notably, selective action at the P13K/Akt and MAP kinase pathways. These pathways regulate survival of transcription factors, proteins that help to switch genes 'on' and 'off', and gene expression (Spencer, 2009).

Flavonoids and their metabolites have been found to cross the blood-brain barrier in areas particularly vulnerable to the effects of again and neurodegeneration, that is, the hippocampus, cerebral cortex, cerebellum and striatum (Datla, Christidou, Widmer, Rooprai, & Dexter, 2001; Milbury & Kalt, 2010; Passamonti, Vrhovsek, Vanzo, & Mattivi, 2005), suggesting that flavonoids exert neuroprotective effects (Nehlig, 2013).

Blood Flow

Cocoa flavanols induce peripheral and vascular blood flow, which may result in the induction of angiogenesis, the formation of new blood vessels and nerve cell growth in the hippocampus (Spencer, 2009). They also display cognitive effects indirectly through their well-established influence on cardiovascular health, including endothelium-dependent vasodilation and increased NO bioavailability, which helps in the maintenance of normal blood flow, to reduce platelet aggregation and inflammation, and improves blood pressure (Flammer et al., 2007; González-Gallego, García-Mediavilla, Sánchez-Campos, & Tuñón, 2014; Grassi, Desideri, & Ferri, 2013; Shrime et al., 2011). This in turn increases cerebral blood flow and blood perfusion throughout the central and peripheral nervous system (Fisher, Sorond, & Hollenberg, 2006; Hollenberg, Fisher, & McCullough, 2009). These outcomes have carry-over effects of cognitive function.

Cognitive Effects Following Acute Administration in Healthy Humans

Several studies have examined the effects of acute cocoa flavanols intake on cognitive functioning in healthy humans, generally resulting in selected cognitive enhancement. The first study to report acute cognitive improvement following cocoa flavanols was a randomised, controlled, double-blinded, balanced and three-period crossover trial of 30 healthy adults who were administered drinks containing 520 mg and 994 mg cocoa flavanols, which was compared to a 46 mg dose (Scholey et al., 2010). Participants completed cognitive testing 90 minutes post-administration. Both of the cocoa flavanols dosages improved performance on a Serial Threes subtraction task while the 994 mg dose also improved reaction time on a Rapid Visual Information Processing (RVIP) working memory vigilance task but reduced accuracy on a Serial Sevens task. In a randomised, single-blinded, counterbalanced and crossover study, 30 participants were administered with dark chocolate containing 720 mg cocoa flavanols and white chocolate containing only trace amounts (Field, Williams, & Butler, 2011). Two hours after administration, participants underwent cognitive testing. Compared to the control condition, the cocoa flavanols improved spatial memory and performance on aspects of a choice reaction time task. In a randomised, double-blind and parallel-groups design study, 72 participants received 0 (placebo), 250 or 500 mg cocoa polyphenols and were assessed on cognitive performance at 1, 2.5 and 4 hours post-dose (Pase et al., 2013), which remained unchanged under each condition and time point.

In a placebo-controlled, double-blind trial, 40 participants administered with 250 mg cocoa resulted in enhanced performance on a Serial Sevens task (Massee et al., 2015). In a small study of 12 male participants who were administered 903 and 15 mg (placebo) cocoa flavanol, cognitive assessments completed 100-minutes post-dose revealed increased cerebral blood oxygenation (NIRS), but there were no effects on performance on the cognitive task.

Conclusion

The research detailed here provides compelling evidence that ginseng, sage and cocoa flavanols each has well-established mechanisms of action that indicate potential cognitive-enhancing properties. While findings from human clinical trials assessing acute administration are somewhat equivocal, it appears that ginseng, sage and cocoa may benefit cognitive functioning in specific cognitive domains. Acute ginseng extract positively effects secondary memory and perhaps, to a smaller extent, attention. Similarly, acute sage administration appears to enhance aspects of memory and lastly, cocoa appears to enhance the performance of various cognitive domains, including memory, attention and executive function. However, it is important to note that this chapter predominately reported significant findings and generally did not report non-significant results.

With the increasing ageing population, it is likely that a growing number of ageing adults will become concerned about their cognitive functioning. While a conventional, pharmaceutical medication targeting cognitive complaints associated with ageing is unavailable, and dietary and herbal supplement use is increasing, it is likely that more people will utilise supplements for cognitive enhancement (Laditka, Laditka, Tait, & Tsulukidze, 2012). However, various issues impede the development and availability of effective nutraceuticals for cognitive functioning. Firstly, there is great variation in the concentration of active constituents in supplements, depending on the plant itself, and preparation and extraction methods used. These factors may contribute to inefficacy of nutraceuticals, as can be illustrated by the finding that ginsenosides content in many brands of ginseng sold in the United States is low to negligible (Tyler & Russo, 2015). Secondly, the legal requirements concerning supplements differ with different legislation governing the claims that can be made about nutraceuticals. A more consistent global legislation would help to improve the quality of available nutraceutical products and the rigour of clinical trials assessing nutraceuticals.

Further research is needed to increase the understanding of synergistic effects of nutraceuticals, which result in complex interactions, and dose- and time-dependent effects (Neale et al., 2013). For example, the effects of multiple nutraceuticals co-administered cannot always be predicted by summating the effects of the individual nutraceuticals. This can be illustrated by the common combination of ginseng and gingko, which, results in better performance on a Serial Sevens task than either of the products taken in isolation (Scholey & Kennedy, 2002). Similarly, within one extract, it is possible for constituents to act synergistically or the effects of a combination of constituents may be less than that of the isolated constituents.

In conclusion there appears to be fairly good evidence that the nutraceutical described above can act as cognitive enhancers. In some cases, such as Salvia, these effects are in keeping with their traditional use. The underlying mechanisms of action are not well understood but may include polypharmacological mechanisms targeting specific neurotransmitter systems and other processes important for cognition (Scholey, 2018).

References

Alkam, T., Nitta, A., Mizoguchi, H., Itoh, A., & Nabeshima, T. (2007). A natural scavenger of peroxynitrites, rosmarinic acid, protects against impairment of memory induced by Aβ25–35. *Behavioural Brain Research, 180*(2), 139–145.

Bailey, R. L., Gahche, J. J., Miller, P. E., Thomas, P. R., & Dwyer, J. T. (2013) Why US adults use dietary supplements. *JAMA Internal Medicine, 173*(5), 355–361.

BCC Research. (2018). Nutraceuticals: Global markets to 2023.

Bell, L., Lamport, D., Butler, L., & Williams, C. (2015). A review of the cognitive effects observed in humans following acute supplementation with flavonoids, and their associated mechanisms of action. *Nutrients, 7*(12), 10290–10306.

Benishin, C. G. (1992). Actions of ginsenoside Rb1 on choline uptake in central cholinergic nerve endings. *Neurochemistry International, 21*(1), 1–5.

Bowling, H., Bhattacharya, A., Klann, E., & Chao, M. V. (2016). Deconstructing brain-derived neurotrophic factor actions in adult brain circuits to bridge an existing informational gap in neuro-cell biology. *Neural Regeneration Research, 11*(3), 363.

Chen, L.-m., Lin, Z.-y., Zhu, Y.-g., Lin, N., Zhang, J., Pan, X.-d., et al. (2012). Ginsenoside Rg1 attenuates β-amyloid generation via suppressing PPARγ-regulated BACE1 activity in N2a-APP695 cells. *European Journal of Pharmacology, 675*(1–3), 15–21.

Chen, L.-M., Zhou, X.-M., Cao, Y.-L., & Hu, W.-X. (2008). Neuroprotection of ginsenoside Re in cerebral ischemia–reperfusion injury in rats. *Journal of Asian Natural Products Research, 10*(5), 439–445.

Chen, Z., Lu, T., Yue, X., Wei, N., Jiang, Y., Chen, M., et al. (2010). Neuroprotective effect of ginsenoside Rb1 on glutamate-induced neurotoxicity: With emphasis on autophagy. *Neuroscience Letters, 482*(3), 264–268.

Datla, K. P., Christidou, M., Widmer, W. W., Rooprai, H. K., & Dexter, D. T. (2001). Tissue distribution and neuroprotective effects of citrus flavonoid tangeretin in a rat model of Parkinson's disease. *Neuroreport, 12*(17), 3871–3875.

Dickinson, A., Blatman, J., El-Dash, N., & Franco, J. C. (2014). Consumer usage and reasons for using dietary supplements: Report of a series of surveys. *Journal of the American College of Nutrition, 33*(2), 176–182.

EFSA Panel on Dietetic Products NaAN. (2012). Scientific opinion on the substantiation of a health claim related to cocoa flavanols and maintenance of normal endothelium-dependent vasodilation pursuant to Article 13(5) of Regulation (EC) No 1924/2006. *EFSA Journal, 10*, 2809.

Ernst, E. (2005). Complementary/alternative medicine for hypertension: A mini-review. *Wiener Medizinische Wochenschrift, 155*(17–18), 386–391.

Field, D. T., Williams, C. M., & Butler, L. T. (2011). Consumption of cocoa flavanols results in an acute improvement in visual and cognitive functions. *Physiology & Behavior, 103*(3–4), 255–260.

Fisher, N. D., Sorond, F. A., & Hollenberg, N. K. (2006). Cocoa flavanols and brain perfusion. *Journal of Cardiovascular Pharmacology, 47*, S210–S214.

Flammer, A. J., Hermann, F., Sudano, I., Spieker, L., Hermann, M., Cooper, K. A., … Corti, R. (2007). Dark chocolate improves coronary vasomotion and reduces platelet reactivity. *Circulation, 116*(21), 2376–2382.

Fllaketov, A., Bogdanova, T., Podvigina, T., & Bodganov, A. (1988). Role of pituitary-adrenocortical system in body adaptation abilities. *Experimental and Clinical Endocrinology & Diabetes, 92*(5), 129–136.

Fonteles, A. A., de Souza, C. M., de Sousa Neves, J. C., Menezes, A. P. F., do Carmo, M. R. S., Fernandes, F. D. P., et al. (2016). Rosmarinic acid prevents against memory deficits in ischemic mice. *Behavioural Brain Research SreeTestContent1, 297*, 91–103.

Foolad, F., & Khodagholi, F. (2013). Dietary supplementation with Salvia sahendica attenuates acetylcholinesterase activity and increases mitochondrial transcription factor A and antioxidant proteins in the hippocampus of amyloid beta-injected rats. *Journal of Pharmacy and Pharmacology, 65*(10), 1555–1562.

Friedl, R., Moeslinger, T., Kopp, B., & Spieckermann, P. G. (2001). Stimulation of nitric oxide synthesis by the aqueous extract of Panax ginseng root in RAW 264.7 cells. *British Journal of Pharmacology, 134*(8), 1663–1670.

Garthwaite, J. (1993). Nitric oxide signalling in the nervous system. *Seminars in Neuroscience, 5*, 171–180.

Garthwaite, J., Charles, S. L., & Chess-Williams, R. (1988). Endothelium-derived relaxing factor release on activation of NMDA receptors suggests role as intercellular messenger in the brain. *Nature, 336*(6197), 385.

González-Gallego, J., García-Mediavilla, M. V., Sánchez-Campos, S., & Tuñón, M. J. (2014). Anti-inflammatory and immunomodulatory properties of dietary flavonoids. *Polyphenols in Human Health and Disease*, 435–452.

Grassi, D., Desideri, G., & Ferri, C. (2013). Protective effects of dark chocolate on endothelial function and diabetes. *Current Opinion in Clinical Nutrition & Metabolic Care, 16*(6), 662–668.

Gu, L., Kelm, M. A., Hammerstone, J. F., Beecher, G., Holden, J., Haytowitz, D., et al. (2004). Concentrations of proanthocyanidins in common foods and estimations of normal consumption. *The Journal of Nutrition, 134*(3), 613–617.

Helms, S. (2004). Cancer prevention and therapeutics: Panax ginseng. *Alternative Medicine Review, 9*(3).

Hiai, S., Yokoyama, H., Oura, H., & Kawashima, Y. (1983). Evaluation of corticosterone secretion-inducing activities of ginsenosides and their prosapogenins and sapogenins. *Chemical and Pharmaceutical Bulletin, 31*(1), 168–174.

Hollenberg, N. K., Fisher, N. D., & McCullough, M. L. (2009). Flavanols, the Kuna, cocoa consumption, and nitric oxide. *Journal of the American Society of Hypertension, 3*(2), 105–112.

Hsieh, M. T., Peng, W. H., Wu, C. R., & Wang, W. H. (2000). The ameliorating effects of the cognitive-enhancing Chinese herbs on scopolamine-induced amnesia in rats. *Phytotherapy Research, 14*(5), 375–377.

Huang, E. J., & Reichardt, L. F. (2001). Neurotrophins: Roles in neuronal development and function. *Annual Review of Neuroscience, 24*(1), 677–736.

Hurst, W. J., Tarka, S. M., Jr., Powis, T. G., Valdez, F., Jr., & Hester, T. R. (2002). Archaeology: Cacao usage by the earliest Maya civilization. *Nature, 418*(6895), 289.

Jeong, C. S. (2002). Effect of butanol fraction of Panax ginseng head on gastric lesion and ulcer. *Archives of Pharmacal Research, 25*(1), 61.

Jiang, P., Li, C., Xiang, Z., & Jiao, B. (2014). Tanshinone IIA reduces the risk of Alzheimer's disease by inhibiting iNOS, MMP-2 and NF-κBp65 transcription and translation in the temporal lobes of rat models of Alzheimer's disease. *Molecular Medicine Reports, 10*(2), 689–694.

Jiang, S., Miao, B., Song, X., & Jiang, Z. (2011). Inactivation of GABAA receptor reduces ginsenoside Rb3 neuroprotection in mouse hippocampal slices after oxygen–glucose deprivation. *Journal of Ethnopharmacology, 133*(2), 914–916.

Jin, S.-H., Park, J.-K., Nam, K.-Y., Park, S.-N., & Jung, N.-P. (1999). Korean red ginseng saponins with low ratios of protopanaxadiol and protopanaxatriol saponin improve scopolamine-induced learning disability and spatial working memory in mice. *Journal of Ethnopharmacology, 66*(2), 123–129.

Jin, X., Liu, P., Yang, F., Zhang, Y.-h., & Miao, D. (2013). Rosmarinic acid ameliorates depressive-like behaviors in a rat model of CUS and Up-regulates BDNF levels in the hippocampus and hippocampal-derived astrocytes. *Neurochemical Research, 38*(9), 1828–1837.

Kennedy, D., Scholey, A., & Wesnes, K. (2001a). Dose dependent changes in cognitive performance and mood following acute administration of Ginseng to healthy young volunteers. *Nutritional Neuroscience, 4*(4), 295–310.

Kennedy, D., Scholey, A., & Wesnes, K. (2001b). Differential, dose dependent changes in cognitive performance following acute administration of a Ginkgo biloba/Panax ginseng combination to healthy young volunteers. *Nutritional Neuroscience, 4*(5), 399–412.

Kennedy, D. O., Dodd, F. L., Robertson, B. C., Okello, E. J., Reay, J. L., Scholey, A. B., et al. (2011). Monoterpenoid extract of sage (Salvia lavandulaefolia) with cholinesterase inhibiting properties improves cognitive performance and mood in healthy adults. *Journal of Psychopharmacology, 25*(8), 1088–1100.

Kennedy, D. O., Pace, S., Haskell, C., Okello, E. J., Milne, A., & Scholey, A. B. (2006). Effects of cholinesterase inhibiting sage (Salvia officinalis) on mood, anxiety and performance on a psychological stressor battery. *Neuropsychopharmacology, 31*(4), 845.

Khodagholi, F., & Ashabi, G. (2013). Dietary supplementation with Salvia sahendica attenuates memory deficits, modulates CREB and its down-stream molecules and decreases apoptosis in amyloid beta-injected rats. *Behavioural Brain Research, 241*, 62–69.

Kiefer, D., & Pantuso, T. (2003). Panax ginseng. *American Family Physician, 68*(8), 1539–1542.

Kim, D. H., Jung, J. S., Suh, H. W., Huh, S. O., Min, S.-K., Son, B. K., et al. (1998). Inhibition of stress-induced plasma corticosterone levels by ginsenosides in mice: Involvement of nitric oxide. *Neuroreport, 9*(10), 2261–2264.

Kim, H.-S., Kang, J.-G., Seong, Y.-H., Nam, K.-Y., & Oh, K.-W. (1995). Blockade by ginseng total saponin of the development of cocaine induced reverse tolerance and dopamine receptor supersensitivity in mice. *Pharmacology Biochemistry and Behavior, 50*(1), 23–27.

Kim, H.-S., Lee, J.-H., Goo, Y.-S., & Nah, S.-Y. (1998). Effects of ginsenosides on Ca2 channels and membrane capacitance in rat adrenal chromaffin cells. *Brain Research Bulletin, 46*(3), 245–251.

Kosaka, K., & Yokoi, T. (2003). Carnosic acid, a component of rosemary (Rosmarinus officinalis L.), promotes synthesis of nerve growth factor in T98G human glioblastoma cells. *Biological and Pharmaceutical Bulletin, 26*(11), 1620–1622.

Kumar, A., Singh, A., & Ekavali. (2015). A review on Alzheimer's disease pathophysiology and its management: An update. *Pharmacological Reports: PR, 67*(2), 195–203.

Laditka, J. N., Laditka, S. B., Tait, E. M., & Tsulukidze, M. M. (2012). Use of dietary supplements for cognitive health: Results of a national survey of adults in the United States. *American Journal of Alzheimer's Disease & Other Dementias®, 27*(1), 55–64.

Lazarus, S. A., Hammerstone, J. F., & Schmitz, H. H. (1999). Chocolate contains additional flavonoids not found in tea. *The Lancet, 354*(9192), 1825.

Lee, Y. W., Kim, D. H., Jeon, S. J., Park, S. J., Kim, J. M., Jung, J. M., et al. (2013). Neuroprotective effects of salvianolic acid B on an Aβ25–35 peptide-induced mouse model of Alzheimer's disease. *European Journal of Pharmacology, 704*(1–3), 70–77.

Li, L., Liu, J., Yan, X., Qin, K., Shi, M., Lin, T., et al. (2011). Protective effects of ginsenoside Rd against okadaic acid-induced neurotoxicity in vivo and in vitro. *Journal of Ethnopharmacology, 138*(1), 135–141.

Liberti, L. E., & Marderosian, A. D. (1978). Evaluation of commercial ginseng products. *Journal of Pharmaceutical Sciences, 67*(10), 1487–1489.

Lopresti, A. L. (2017). Salvia (Sage): A review of its potential cognitive-enhancing and protective effects. *Drugs in R&D, 17*(1), 53–64.

Lu, T., Jiang, Y., Zhou, Z., Yue, X., Wei, N., Chen, Z., et al. (2011). Intranasal ginsenoside Rb1 targets the brain and ameliorates cerebral ischemia/reperfusion injury in rats. *Biological and Pharmaceutical Bulletin, 34*(8), 1319–1324.

Marcelo, F., Dias, C., Martins, A., Madeira, P. J., Jorge, T., Florêncio, M. H., et al. (2013). Molecular recognition of rosmarinic acid from Salvia sclareoides extracts by acetylcholinesterase: A new binding site detected by NMR spectroscopy. *Chemistry–A European Journal, 19*(21), 6641–6649.

Marinac, J. S., Buchinger, C. L., Godfrey, L. A., Wooten, J. M., Sun, C., & Willsie, S. K. (2007). Herbal products and dietary supplements: A survey of use, attitudes, and knowledge among older adults. *The Journal of the American Osteopathic Association, 107*(1), 13–23.

Massee, L. A., Ried, K., Pase, M., Travica, N., Yoganathan, J., Scholey, A., et al. (2015). The acute and sub-chronic effects of cocoa flavanols on mood, cognitive and cardiovascular health in young healthy adults: A randomized, controlled trial. *Frontiers in Pharmacology, 6*, 93.

Merad, M., Soufi, W., Ghalem, S., Boukli, F., Baig, M., Ahmad, K., et al. (2014). Molecular interaction of acetylcholinesterase with carnosic acid derivatives: A neuroinformatics study. *CNS & Neurological Disorders-Drug Targets (Formerly Current Drug Targets-CNS & Neurological Disorders), 13*(3), 440–446.

Milbury, P. E., & Kalt, W. (2010). Xenobiotic metabolism and berry flavonoid transport across the blood–brain barrier. *Journal of Agricultural and Food Chemistry, 58*(7), 3950–3956.

Mogil, J. S., Shin, Y.-H., McCleskey, E. W., Kim, S.-C., & Nah, S.-Y. (1998). Ginsenoside Rf, a trace component of ginseng root, produces antinociception in mice. *Brain Research, 792*(2), 218–228.

More, S. V., Kumar, H., Cho, D.-Y., Yun, Y.-S., & Choi, D.-K. (2016). Toxin-induced experimental models of learning and memory impairment. *International Journal of Molecular Sciences, 17*(9), 1447.

Moss, L., Rouse, M., Wesnes, K. A., & Moss, M. (2010). Differential effects of the aromas of Salvia species on memory and mood. *Human Psychopharmacology: Clinical and Experimental, 25*(5), 388–396.

Neale, C., Camfield, D., Reay, J., Stough, C., & Scholey, A. (2013). Cognitive effects of two nutraceuticals Ginseng and Bacopa benchmarked against modafinil: A review and comparison of effect sizes. *British Journal of Clinical Pharmacology, 75*(3), 728–737.

Nehlig, A. (2013). The neuroprotective effects of cocoa flavanol and its influence on cognitive performance. *British Journal of Clinical Pharmacology, 75*(3), 716–727.

Nitta, H., Matsumoto, K., Shimizu, M., Ni, X., & Watanabe, H. (1995). Panax ginseng extract improves the scopolamine-induced disruption of 8-arm radial maze performance in rats. *Biological and Pharmaceutical Bulletin, 18*(10), 1439–1442.

Pase, M. P., Scholey, A. B., Pipingas, A., Kras, M., Nolidin, K., Gibbs, A., et al. (2013). Cocoa polyphenols enhance positive mood states but not cognitive performance: A randomized, placebo-controlled trial. *Journal of Psychopharmacology, 27*(5), 451–458.

Passamonti, S., Vrhovsek, U., Vanzo, A., & Mattivi, F. (2005). Fast access of some grape pigments to the brain. *Journal of Agricultural and Food Chemistry, 53*(18), 7029–7034.

Patil, C. S., Singh, V. P., Satyanarayan, P., Jain, N. K., Singh, A., & Kulkarni, S. K. (2003). Protective effect of flavonoids against aging-and lipopolysaccharide-induced cognitive impairment in mice. *Pharmacology, 69*(2), 59–67.

Petkov, V. (1978). Effect of ginseng on the brain biogenic monoamines and 3′, 5′-AMP system. Experiments on rats. *Arzneimittel-Forschung, 28*(3), 388.

Prince, M., Bryce, R., Albanese, E., Wimo, A., Ribeiro, W., & Ferri, C. P. (2013). The global prevalence of dementia: A systematic review and metaanalysis. *Alzheimer's & Dementia, 9*(1), 63–75. e2.

Rasoolijazi, H., Azad, N., Joghataei, M., Kerdari, M., Nikbakht, F., & Soleimani, M. (2013). The protective role of carnosic acid against beta-amyloid toxicity in rats. *The Scientific World Journal, 2013.*

Repantis, D., Schlattmann, P., Laisney, O., & Heuser, I. (2010). Modafinil and methylphenidate for neuroenhancement in healthy individuals: A systematic review. *Pharmacological Research, 62*(3), 187–206.

Sala, F., Mulet, J., Choi, S., Jung, S.-Y., Nah, S.-Y., Rhim, H., et al. (2002). Effects of ginsenoside Rg2 on human neuronal nicotinic acetylcholine receptors. *Journal of Pharmacology and Experimental Therapeutics, 301*(3), 1052–1059.

Salim, K. N., McEwen, B. S., & Chao, H. M. (1997). Ginsenoside Rb1 regulates ChAT, NGF and trkA mRNA expression in the rat brain. *Molecular Brain Research, 47*(1–2), 177–182.

Sallam, A., Mira, A., Ashour, A., & Shimizu, K. (2016). Acetylcholine esterase inhibitors and melanin synthesis inhibitors from Salvia officinalis. *Phytomedicine, 23*(10), 1005–1011.

Schneider, W. J., & McGrew, K. S. (2018). The Cattell–Horn–Carroll theory of cognitive abilities. In D. P. Flanagan & E. M. McDonough (Eds.), *Contemporary intellectual assessment: Theories, tests, and issues* (pp. 73–163). The Guilford Press.

Scholey, A. (2018). Nutrients for neurocognition in health and disease: Measures, methodologies and mechanisms. *Proceedings of the Nutrition Society, 77*(1), 73–83.

Scholey, A. B., French, S. J., Morris, P. J., Kennedy, D. O., Milne, A. L., & Haskell, C. F. (2010). Consumption of cocoa flavanols results in acute improvements in mood and cognitive performance during sustained mental effort. *Journal of Psychopharmacology, 24*(10), 1505–1514.

Scholey, A. B., & Kennedy, D. O. (2002). Acute, dose-dependent cognitive effects of Ginkgo biloba, Panax ginseng and their combination in healthy young volunteers: Differential interactions with cognitive demand. *Human Psychopharmacology: Clinical and Experimental, 17*(1), 35–44.

Scholey, A. B., Tildesley, N. T., Ballard, C. G., Wesnes, K. A., Tasker, A., Perry, E. K., et al. (2008). An extract of Salvia (sage) with anticholinesterase properties improves memory and attention in healthy older volunteers. *Psychopharmacology, 198*(1), 127–139.

Scott, G. I., Colligan, P. B., Ren, B. H., & Ren, J. (2001). Ginsenosides Rb1 and Re decrease cardiac contraction in adult rat ventricular myocytes: Role of nitric oxide. *British Journal of Pharmacology, 134*(6), 1159–1165.

Seo, Y.-J., Kwon, M.-S., Choi, H.-W., Jang, J.-E., Lee, J.-K., Sun, Y., et al. (2008). Intracerebroventricular gisenosides are antinociceptive in proinflammatory cytokine-induced pain behaviors of mice. *Archives of Pharmacal Research, 31*(3), 364–369.

Shekarchi, M., Hajimehdipoor, H., Saeidnia, S., Gohari, A. R., & Hamedani, M. P. (2012). Comparative study of rosmarinic acid content in some plants of Labiatae family. *Pharmacognosy Magazine, 8*(29), 37.

Shrime, M. G., Bauer, S. R., McDonald, A. C., Chowdhury, N. H., Coltart, C. E., & Ding, E. L. (2011). Flavonoid-rich cocoa consumption affects multiple cardiovascular risk factors in a meta-analysis of short-term studies. *The Journal of Nutrition, 141*(11), 1982–1988.

Sloley, B. D., Pang, P., Huang, B.-H., Ba, F., Li, F. L., Benishin, C. G., et al. (1999). American ginseng extract reduces scopolamine-induced amnesia in a spatial learning task. *Journal of Psychiatry and Neuroscience, 24*(5), 442.

Smach, M., Hafsa, J., Charfeddine, B., Dridi, H., & Limem, K. (2015). Effects of sage extract on memory performance in mice and acetylcholinesterase activity. *Annales Pharmaceutiques Francaises, 73*(4), 281–288.

Spencer, J. P. (2009). Flavonoids and brain health: Multiple effects underpinned by common mechanisms. *Genes & Nutrition, 4*(4), 243.

Sünram-Lea, S., Birchall, R., Wesnes, K., & Petrini, O. (2005). The effect of acute administration of 400 mg of Panax ginseng on cognitive performance and mood in healthy young volunteers. *Current Topics in Nutraceutical Research, 3*(1), 65–74.

Tachikawa, E., Kudo, K., Harada, K., Kashimoto, T., Miyate, Y., Kakizaki, A., et al. (1999). Effects of ginseng saponins on responses induced by various receptor stimuli. *European Journal of Pharmacology, 369*(1), 23–32.

Takeda, H., Tsuji, M., Yamada, T., Masuya, J., Matsushita, K., Tahara, M., et al. (2006). Caffeic acid attenuates the decrease in cortical BDNF mRNA expression induced by exposure to forced swimming stress in mice. *European Journal of Pharmacology, 534*(1–3), 115–121.

Teng, Y., Zhang, M.-Q., Wang, W., Liu, L.-T., Zhou, L.-M., Miao, S.-K., et al. (2014). Compound danshen tablet ameliorated aβ 25–35-induced spatial memory impairment in mice via rescuing imbalance between cytokines and neurotrophins. *BMC Complementary and Alternative Medicine, 14*(1), 23.

Tildesley, N. T., Kennedy, D. O., Perry, E. K., Ballard, C. G., Savelev, S., Wesnes, K. A., et al. (2003). Salvia lavandulaefolia (Spanish sage) enhances memory in healthy young volunteers. *Pharmacology Biochemistry and Behavior, 75*(3), 669–674.

Toda, N., Ayajiki, K., Fujioka, H., & Okamura, T. (2001). Ginsenoside potentiates NO-mediated neurogenic vasodilatation of monkey cerebral arteries. *Journal of Ethnopharmacology, 76*(1), 109–113.

Tyler, V. M., & Russo, E. B. (2015). *Handbook of psychotropic herbs: A scientific analysis of herbal remedies for psychiatric conditions*. Routledge.

Wang, H., Abajobir, A. A., Abate, K. H., Abbafati, C., Abbas, K. M., Abd-Allah, F., et al. (2017). Global, regional, and national under-5 mortality, adult mortality, age-specific mortality, and life expectancy, 1970–2016: A systematic analysis for the Global Burden of Disease Study 2016. *The Lancet, 390*(10100), 1084–1150.

Wang, W., Huang, C.-Y., Tsai, F.-J., Tsai, C.-C., Yao, C.-H., & Chen, Y.-S. (2011). Growth-promoting effects of quercetin on peripheral nerves in rats. *The International Journal of Artificial Organs, 34*(11), 1095–1105.

Wilmoth, J. R. (2000). Demography of longevity: Past, present, and future trends. *Experimental Gerontology, 35*(9–10), 1111–1129.

Xu, S. L., Bi, C. W., Choi, R. C., Zhu, K. Y., Miernisha, A., Dong, T. T., et al. (2013). Flavonoids induce the synthesis and secretion of neurotrophic factors in cultured rat astrocytes: A signaling response mediated by estrogen receptor. *Evidence-Based Complementary and Alternative Medicine, 2013*.

Ye, R., Li, N., Han, J., Kong, X., Cao, R., Rao, Z., et al. (2009). Neuroprotective effects of ginsenoside Rd against oxygen-glucose deprivation in cultured hippocampal neurons. *Neuroscience Research, 64*(3), 306–310.

Ye, R., Yang, Q., Kong, X., Han, J., Zhang, X., Zhang, Y., et al. (2011). Ginsenoside Rd attenuates early oxidative damage and sequential inflammatory response after transient focal ischemia in rats. *Neurochemistry International, 58*(3), 391–398.

Yun, T.-K. (2001). Panax ginseng—A non-organ-specific cancer preventive? *The Lancet Oncology, 2*(1), 49–55.

Zhang, C., Du, F., Shi, M., Ye, R., Cheng, H., Han, J., et al. (2012). Ginsenoside Rd protects neurons against glutamate-induced excitotoxicity by inhibiting Ca 2+ influx. *Cellular and Molecular Neurobiology, 32*(1), 121–128.

Zhang, J., Qu, Z., Liu, Y., & Deng, H. (1990). Preliminary study on antiamnestic mechanism of ginsenoside Rg1 and Rb1. *Chinese Medical Journal, 103*(11), 932–938.

Zhao, Y., Xu, P., Hu, S., Du, L., Xu, Z., Zhang, H., et al. (2015). Tanshinone II A, a multiple target neuroprotectant, promotes caveolae-dependent neuronal differentiation. *European Journal of Pharmacology, 765*, 437–446.

Zhu, J.-r., Tao, Y.-f., Lou, S., & Wu, Z.-m. (2010). Protective effects of ginsenoside Rb 3 on oxygen and glucose deprivation-induced ischemic injury in PC12 cells. *Acta Pharmacologica Sinica, 31*(3), 273.

4

Nutritional Interventions to Improve Cognitive Function

Steven Roodenrys

The ingestion of foods, drinks, or supplements with the specific aim of improving cognitive function as a practice is both ancient and ubiquitous. For millions of people around the world, the ritual of a cup of coffee early in the day is associated with a subjective increase in alertness and improved cognition. Many people voice their belief that they "need a coffee to get going" or that they are "hopeless" until they have had the first of the day. It is so commonplace that it might seem trivial to mention coffee drinking, but overuse and the deliberate use of caffeine as a stimulant to enhance or maintain performance are widespread and may have more serious consequences.

At the same time, the deliberate use of less common foodstuffs or herbal preparations, particularly in order to ameliorate a deficit in cognitive function, has a long history. In traditional medicine practices such as Ayurveda and traditional Chinese medicine, there is a long history of

S. Roodenrys (✉)
School of Psychology, University of Wollongong, Wollongong, NSW, Australia
e-mail: steven_roodenrys@uow.edu.au

© The Author(s) 2021
M. Hall et al. (eds.), *Chemically Modified Minds*,
https://doi.org/10.1007/978-981-15-6771-1_4

plant- or animal-derived supplements to improve memory and other cognitive functions. Interestingly, in one search for memory-improving substances in a compendium of traditional Chinese medicinal texts written before 1950 (May, Lu, Lu, Zhang, & Xue, 2013), Ginkgo Biloba, which has arguably been adopted most widely in Western society as a memory enhancer, does not appear. A recent meta-analysis of 13 randomised controlled trials of Ginkgo supplementation in healthy adults found no evidence that it improved memory (Laws, Sweetnam, & Kondel, 2012). Together, this highlights two issues that distinguish supplements as cognitive enhancers from pharmaceuticals: the vagary of the processes that bring particular supplements to prominence and the degree to which evidence often fails to justify popular usage.

There are hundreds, if not thousands, of websites either selling or promoting foods or supplements as cognitive enhancers, or for brain function.[1] The psychological processes that contribute to people buying treatments that have little or no evidence to support their use are interesting fields of study in themselves, but it is clear that improved cognitive function is something that people are prepared to pay for. For the sake of simplicity, in the rest of this chapter I will refer to taking a dietary supplement such as herbs or consuming food or drink for the purpose of improving health as a dietary supplement, and I will limit the discussion to studies with measures of cognitive function in humans. Foods that have been identified as having a specific health-related effect, or have been altered to do so, are also referred to as functional foods in academic literature and increasingly in literature aimed at the public.

[1] It should be noted that cognitive enhancement and brain function are often used interchangeably in promotion to the public, but they are not synonymous in practice or evidence. It is quite possible to find evidence for an influence of a substance on in vivo brain activity or at the cell level in *in vitro* studies, but these do not necessarily translate into a measurable effect on cognitive performance. Cellular studies are often done with supra-physiological concentrations that cannot be achieved through diet (Vauzour, 2014).

How Many People Use Dietary Supplements for Cognition, and Why?

There is very little research addressing the question of how many people use dietary supplements to enhance cognitive function, although there are large health survey studies that asked about general dietary supplementation from a number of countries that suggest supplement use is pervasive. For example, Bailey, Gahche, Miller, Thomas, and Dwyer (2013) reported data from almost 12,000 interviews in the National Health and Nutrition Examination Survey in the United States. Almost half of the respondents had used a dietary supplement in the previous month, and it is worth noting that the questions refer to "vitamins, minerals, herbals and other dietary supplements" so may well not capture the consumption of specific foods for health reasons, so the proportion may be larger. In Britain, a similar survey found the proportion may be lower, but still substantial, with 33% of 19–64-year-olds reporting to have taken a supplement in the previous year and 40% in those aged 65 and older (Bates, Lennox, Prentice, Bates, & Swan, 2011). Data from over 19,000 participants in the Australian National Health Survey found 43% had used a supplement in the previous two weeks (O'Brien, Malacova, Sherriff, & Black, 2017).

The trend for greater usage of dietary supplements by older adults appears to be consistent across different countries (e.g. Bailey et al., 2013; Kofoed, Christensen, Dragsted, Tjønneland, & Roswall, 2015; Reinert, Rohrmann, Becker, & Linseisen, 2007) and no doubt reflects the increase in health problems with age. As will be discussed later, one of the areas of interest in relation to diet and cognition is in preventing or delaying the decline in cognitive function associated with age and the risk of dementia, so it seems likely that this partly reflects a greater interest in maintaining general health as people age. A stronger interest in health appears to drive the behaviour, with females being more likely to take supplements than males, and supplement users are more likely to be a healthy weight, exercising, and not smoking (Bailey et al., 2013; Reinert et al., 2007). At the same time, some health issues are particularly associated with supplement use. Brownie (2006) reported that people with musculo-skeletal disorders like arthritis and osteoporosis were twice as likely to use supplements as those without this problem. de Jong, Ocké, Branderhorst, and Friele (2003) reported the results of a study of Dutch supplement and functional food use and found some variation in the association of demographic and lifestyle factors with different supplements/foods. This suggests that some people do use particular supplements for a specific purpose. This is consistent with recent survey data from Dickinson, Blatman, El-Dash, and Franco (2014) showing that although the most commonly used supplement is multivitamins and the most commonly cited reasons for taking a supplement are overall health and wellness, and to fill a nutrient gap, many more specific health issues were cited. In this research, over five annual surveys, roughly two-thirds of the samples reported using supplements and 13% of respondents endorsed "mental focus, concentration" as a reason for supplement use while less than 10% reported use for memory function. This suggests as much as 10% of the population may be using supplements to improve or maintain cognitive function.

Although the Dickinson et al. (2014) study included concentration and memory in a list of 28 potential reasons for taking supplements presented to participants, most studies have not asked participants about cognitive function, unless it was the focus of the study, for example, a survey of stimulant use in students. Some of these studies also include supplements in the list of substances to improve cognition that participants are asked about. It seems that the major reasons people take supplements to improve cognitive function are in order to perform better in an academic setting when young and to maintain cognitive function when old.

Leaving aside the issue of how people might consider the evidence for the benefits of a dietary supplement, there are other psychological factors that influence their use. Cox, Koster, and Russell (2004) investigated the perceptions of middle-aged adults in relation to a number of possible supplements to improve memory function, conducted within the framework of Protection Motivation Theory. Participants were given texts describing how different foods or supplements were being developed to improve memory, and provided information about taste and function. They were asked to rate their intention to consume the food or supplement and a number of other questions. The results showed intention to use the supplement was greatest if it was a natural food, lower for a non-food supplement (e.g. a pill), lower for a food artificially sweetened to offset bitterness, and lowest for a genetically modified food that contained double the content of the active constituent. Not surprisingly, perceived efficacy of the supplement and perceived ability to consume the supplement were the strongest predictors of intention to use the supplement. In addition, the individual's perceived severity of the threat of memory loss and the importance of being vulnerable to memory loss were significant predictors.

Cardello and Schutz (2003) asked respondents about intention to take a supplement for a small number of benefits, which included improved thinking. They found that taste, the type of benefit, the source of the information, and the frequency of use required were all rated as more important than the form of the supplement, but the preferred form was as a capsule/tablet rather than a drink or solid food. This suggests that convenience is a factor affecting decisions regarding supplement use. In a study of supplement use for any purpose, Denham (2017) analysed data from a large US survey to look at the influence of personality factors and locus of control on self-reported usage. Locus of control refers to the degree to which individuals view the outcomes or events they experience as being within their control, so someone with an internal locus of control will see their health outcomes as largely within their own control. While none of the standard personality factors were independent predictors of supplement use, the degree of agreement with the internal locus of control items was positively related to supplement use. However, so was the response to the single item measuring external locus of control. Further analyses suggested that the relationship between supplement use and external locus of control was driven by increased external locus of control in older individuals, accompanied by increased use of supplements. It seems that age encourages the belief that our health is outside of our control but also the increase in actual health problems or perceived risk of problems encourages supplement use.

Before discussing evidence for the use of dietary supplements by students seeking to improve cognitive function, it is worth briefly mentioning the research on the illicit use of prescription drugs for this purpose.

There is a growing body of research on the prevalence and reasons for students using drugs which include opioids and tranquillizers, but those including stimulants are relevant to cognitive enhancement. A survey of a large sample of students from a single US university found a lifetime prevalence of use of 8% (Teter, McCabe, LaGrange, Cranford, & Boyd, 2006) which matched national data (7%, McCabe, Knight, Teter, & Wechsler, 2005) and the most commonly used substance was an amphetamine-dextroamphetamine combination prescribed for attention deficit disorder, presumably because of its availability. Three of the reasons listed on the survey received two-thirds of the endorsements; for concentration, to help with study, and to increase alertness, all of which can be seen as aimed at improving learning. A survey of several hundred university students in Britain found lifetime prevalence between 3% and 8% for the three stimulants investigated, and the major reason for use was to enhance cognition, followed by offset sleep deprivation (Singh, Bard, & Jackson, 2014), while a survey of over 4000 students across Europe reported a prevalence rate of 6% (Helmer et al., 2016). A recent meta-analysis of studies of the motives for illicit prescription drug use in students also found that academic outcomes and staying awake were the major reasons for use (Bennett & Holloway, 2017).

There are no studies that examine the use of a range of dietary supplements by students to enhance cognition, although some studies of stimulants have included caffeine in the research. Singh, Bard, and Jackson (2014) reported a lifetime prevalence of caffeine pill use in their study of university students of almost 50%. While 60% of users had done so to offset sleep deprivation, half of them also endorsed using it to enhance cognition. A survey of German students found a lifetime prevalence for the purpose of cognitive enhancement of 53% for coffee, 39% for caffeinated drinks, and 10% for caffeine pills (Franke et al., 2011).

We appear to know very little about student use of supplements specifically to enhance cognition. Although there is considerable literature on college student-athlete use of dietary supplements, there are relatively few on a general student population, and a number have used only nutrition students. A recent study by Lieberman et al. (2015) of over 1200 students at five different universities in the United States asked about supplement use according to a definition that does not include whole foods. They found that two-thirds used a supplement at least once a week. Caffeine was not specifically listed in the survey, but 16% of the students listed it as something they had taken as a supplement in the previous six months, presumably to increase alertness. Furthermore, 8% consumed fish oil which seems more likely to be for cognitive function than cardiovascular health in this age group. Of those who used any supplement, 19% endorsed "performance enhancement" as a reason for use, but this does not distinguish between cognitive enhancement for academic outcomes and other types of performance, such as athletic.

Kobayashi, Sato, Umegaki, and Chiba (2017) used an internet research company to survey over 9000 Japanese college students and found 32% were using or had used supplements. Their definition of a supplement included whole foods but not foods consumed daily. While over half of those taking supplements said it was to improve their health, the survey did not include cognitive enhancement as a reason for use and less than 2% of the responded with the "other" option, suggesting that the use of dietary supplements for cognitive enhancement is very low.

Dietary Cognitive Enhancers with Immediate Effects

As indicated above, caffeine, in either coffee, or energy drinks, or pills, is widely used to improve concentration, alertness, and counteract the effects of sleep deprivation. There has been some debate, however, as to whether the effect of caffeine is more to do with reversing withdrawal effects (needing that first coffee in the morning to get going) than a genuine boost to baseline performance. The effect observed in studies varies depending on whether subjects have been deprived before the study, how much caffeine they habitually consume, and the cognitive function being tested (for a review, see Nehlig, 2010). Nehlig (2010) actually concludes that caffeine is not a cognitive enhancer, using a narrow definition of cognition as memory and higher-order processes like decision-making. Caffeine does have an effect on arousal, and this has an impact on performance in tasks requiring sustained attention and increases the speed of responding.

Another substance consumed for the potential immediate effect on cognition is sugar, or glucose. Glucose is the energy source for our cells and effects of variation in glucose levels on cognition are well known in areas such as diabetes, and in relation to children's diets, particularly the impact of not eating breakfast. However, a dose of glucose over and above normal eating patterns has been demonstrated to enhance performance in some cognitive tasks. There is wide variation in methods and cognitive tasks used in the literature, and although a meta-analysis by Riby (2004) found an average effect size of glucose administration of 0.56, this was moderated by cognitive domain and by whether the participants had fasted or not. In unfasted groups the effect size was 0.19, while fasted groups showed an effect of 0.71. The cognitive domains of episodic and working memory showed the largest effects. The magnitude of the effect was also affected by dose with a smaller effect size for doses greater than 25 g than doses of 25 g. This pattern suggests that glucose has an effect that is fairly specific to memory, but it is relatively small. As episodic memory tests involve presentation and retrieval of the information a short time later, the research does not provide strong support for the practice of some students to take sweets into exams, and effects in general are likely to be small if students have a normal diet and glucose function.

In relation to herbal supplements, one recent review (Neale, Camfield, Reay, Stough, & Scholey, 2012) concluded that there are acute effects of ginseng, another concluded that there was insufficient evidence to justify the claim that ginseng produces an acute effect (Smith, Williamson, Putnam, Farrimond, & Whalley, 2014). A number of studies have reported significant effects of a single dose of ginkgo on cognitive function, however the pattern of effects overall is inconsistent. For example, studies from the same research group, using the same dose and same battery of cognitive tasks (Kennedy, Scholey, & Wesnes, 2000, 2002), have reported different patterns of significant effects over tasks and time since administration. On the whole, the evidence for an acute effect of any dietary supplement, other than caffeine and glucose, is unconvincing.

Longer-Term Effects in Healthy Adults

The discussion of effects of chronic consumption of dietary supplements on cognitive function is here divided into those looking at healthy adults and those looking at slowing or reversing cognitive decline with age. This recognises the two distinct motivations for supplement use, to improve current function, or simply to maintain it in the face of what seems like inevitable decline. To pre-empt the discussion, and generalising the earlier statements about Gingko Biloba, there is modest evidence that any supplement improves cognitive function in healthy people, whereas there is evidence that cognitive decline is affected by diet and can be altered by supplement use. This pattern suggests that in the absence of pathology, a healthy varied diet provides sufficient nutrients for effective brain function, and supplementation may not be able to improve on our typical level of cognitive function.

As stated earlier, in a meta-analysis of the 13 studies that met their criteria, Laws et al. (2012) found no evidence that Ginkgo Biloba improved memory, attention, or executive function with the effect sizes being close to zero. Although one study (Sathyanarayanan et al., 2013) of *Bacopa monnieri* (Brahmi) recently found no effect of 12 weeks of supplementation in healthy adults, a review including six studies of the same duration concluded that there was some evidence that it does have an effect on memory but not on other cognitive functions (Pase et al., 2012). The most consistent result is an effect on memory after a delay rather than immediate recall, although not for semantically coherent material, such as a short story. Most studies focus on laboratory tasks; however, while we (Roodenrys et al., 2002) found a significant effect in a laboratory task, we also assessed subjective memory function and found no effect. It may be that the effect found on well-controlled laboratory tasks is not sufficient to make a practical difference to healthy adults. One problem with intervention studies is that they often do not assess the adequacy of the participants' diet, so it may be that any effect that occurs is restricted to those who are deficient in some aspect of their diet or have impaired memory function.

Diet and Cognitive Decline in Older Adults

There has been considerable research interest in the role of diet in cognitive function as we age. Epidemiological studies focus on variations in typical diet and how they relate to cognitive function, with both negative and positive effects being researched, and also on the relationship between variation in the consumption of specific foods and cognitive function. Intervention studies have looked at the effects of supplementing the participants' normal diet. It seems that the impact of dietary interventions may depend on the type of intervention and the level of cognitive function. As outlined above, supplementation in healthy adults does not have much impact on cognitive function, and there are relatively few long-term studies of supplementation tracking any decline in cognitive function. However, global changes of diet may impact the rate of decline as there are well-researched effects of habitual dietary patterns on physical health and levels of cognitive function. There are also a growing number of studies of specific supplements showing improvements in cognitive function, or a slowing in the rate of decline, in groups with pathological conditions, such as dementia.

Epidemiological studies show that some patterns of diet are beneficial for general health. The Mediterranean diet (MeDi) is one that is high in vegetables, legumes, fruits, unrefined grains, fish, and olive oil, and has been argued to reduce the likelihood of Alzheimer's disease. The outcomes of studies looking at the relationship of the MeDi to cognitive function have been mixed. Loughrey, Lavecchia, Brennan, Lawlor, and Kelly (2017) recently published a meta-analysis of the literature and concluded there is an effect of the MeDi on global cognition. Singh, Parsaik et al. (2014) reported a meta-analysis of studies examining the risk of developing Mild Cognitive Impairment or Alzheimer's and found that the MeDi did reduce the risk of developing these pathological conditions.

Morris and her colleagues have adapted the scores for adherence to a MeDi and a blood pressure-lowering diet to also take account of recent work suggesting beneficial effects of berries and leafy vegetables in the diet, and termed this pattern the MIND diet. Morris et al. (2015a) found that adherence to the MIND diet reduced the chance of developing

Alzheimer's over a 4.5 year period. In the same cohort study, they found that adherence to the diet lowered the rate of cognitive decline over a 10 year period (Morris et al., 2015b).

The benefits of better cognitive function in old age are not just psychological. Dementia has an enormous economic and social impact as sufferers require more care and assistance. There is good evidence that the assistance required is directly linked to cognitive function, in that cognitive decline has been shown to precede functional decline, but the relationship may well be reciprocal, as social engagement is also associated with better cognitive function. In a comprehensive review of factors associated with functional outcomes, Stuck et al. (1999) concluded there was strong evidence for an association between cognitive function and disability in older people. In addition, it has been shown that the level of cognitive function is related to the subsequent rate of decline in basic care abilities (e.g. Atchison, Massman, & Doody, 2011; Helvik et al., 2015).

Not all cognitive functions appear to be equally important for daily living. Johnson, Lui, and Yaffe (2007) found that executive function was a better predictor than a global cognition score of current and subsequent ability to cope with activities of daily living. Helmes and Klinger (2017) recently found that various domains of cognitive function were differentially related to a variety of activities of daily living, with memory being consistently related to all activities as well as visuospatial processing (assessed by a task requiring manipulation of objects). Dietary supplements that can delay cognitive decline will also have an impact on daily function, but some supplements may prove to be more beneficial than others as a consequence of which cognitive functions are affected.

As mentioned above, the relationship between activities of daily living and cognition is reciprocal. A reduced ability to engage in social activities will have a negative impact on cognitive function. A number of studies have shown that social isolation predicts cognitive decline (for a review, see Cacioppo & Hawkley, 2009) but specific psychological factors appear to be important. Recently, DiNapoli, Wu, and Scogin (2014) found that perceived isolation, in terms of lack of support and loneliness, explained more variation in current level of cognitive function than social disconnectedness, defined as having a small social network and less social activity. However, the impact of cognitive decline on functional ability may be larger than the impact of functional capacity on cognition. In a longitudinal study with four measurements over five years, Zahodne, Manly, MacKay-Brandt, and Stern (2013) found that cognitive scores were stronger predictors of subsequent functional ability than functional scores were of subsequent cognitive performance.

Whole Foods as Cognitive Enhancers

There is a growing interest in the use of foods, in a more or less unprocessed form, as a means of enhancing cognition, and it is becoming a very active area of research. As mentioned above, the MIND diet was developed to take into account recent evidence about the effects on cognitive function of particular foods as part of the diet. The focus of much of this research is on a class of chemicals called flavonoids which can be broken into a number of sub-classes, including anthocyanins, flavanols, and flavonols. The research to date has large gaps in epidemiological evidence, acute and long-term intervention studies for various flavonoids making it impossible to draw broad conclusions.

Antioxidant action was initially suggested as the possible mechanism for many cognitive enhancers, including foods. However, documented effects on cardiovascular health (for a review, see Wang, Ouyang, Liu, & Zhao, 2014) mean it is also possible that any effect on cognition is through an effect on vascular health and blood flow to the brain. In addition, it is possible that specific flavonoids could affect some neurotransmitters and improve cognition as a result (see Rendeiro, Rhodes, & Spencer, 2015, for a review of mechanisms by which flavonoids might influence brain function).

There is good evidence for an effect of long-term consumption of flavonoids on cognitive function, but the situation is complicated by variation across studies in what flavonoids have been measured, particularly in the longitudinal studies which have often been aimed at cardiovascular function and included dietary measures that were not targeted at measuring flavonoids in the diet. A number of studies have found higher flavonoid intake was associated with better cognitive function and a slower rate of decline (e.g. Devore, Kang, Breteler, & Grodstein, 2012;

Letenneur, Proust-Lima, Le Gouge, Dartigues, & Barberger-Gateau, 2007; Nurk et al., 2009; Root, Ravine & Harper, 2015). Although, Nooyens et al. (2015) found no relationship between flavonoid intake and global cognition, the average age of their participants was 55 at baseline, and less than 10% of them would have been over 70 at the final test. It may be that the positive effects of flavonoid intake are only seen in older adults.

Although flavonoid intake may help cognition in older adults, it appears to be unrelated to the risk of developing dementia. This null effect has been reported from analyses of data from the Rotterdam study at different time points (Devore et al., 2010; Engelhart et al., 2002) and other studies (e.g. Commenges et al., 2000; Laurin, Masaki, Foley, White, & Launer, 2004). It suggests that the effect on cognitive decline is through a mechanism other than that responsible for Alzheimer's dementia.

Although there are a number of intervention studies that report positive effects of supplementation with specific foods, they do not yet provide convincing evidence of an effect. The studies tend to have small numbers of participants. They often test many cognitive functions and find only one or two show an effect, and control over the type I error rate is arguably insufficient. However, the main concern is a lack of consistent replication of effects. There appears to be no attempt to replicate some findings, as yet, and other results have not been replicated. For example, Krikorian, Nash, Shidler, Shukitt-Hale, and Joseph (2010) found an effect of anthocyanin-rich Concord grape juice on learning and a trend on memory measures from the auditory verbal learning task in 12 elderly people with mild cognitive decline. However, a second study with 21 similar participants and the same dosage over 16 weeks did not replicate the effect on learning or memory (Krikorian et al., 2012). In a similar study with anthocyanin-rich cherry juice and older adults with a diagnosis of mild dementia, Kent et al. (2017) did find a significant improvement on several measures of memory function with this task, but more studies are needed before conclusions can be drawn.

Flavonols are found in a variety of foods, such as chocolate and apples, but the most common source for many people is from drinking tea. Studies on the effect of cocoa flavanols have found a beneficial effect on executive function in healthy adults (Mastroiacovo et al., 2015) and people with Mild Cognitive Impairment (Desideri et al., 2012). However, Sorond, Hurwitz, Salat, Greve, and Fisher (2013) found an effect on executive tasks only in their elderly participants who had an impaired vascular response to neuronal activity, indicative of pathology. Overall, the studies suggest there may be an effect of long-term cocoa consumption on executive function, but again, more studies are needed.

Conclusion

There is a strong need for more research on the possibility of enhancing cognition through dietary supplements. Cognitive function can be enhanced immediately by some substances but this is largely due to an increase in alertness and the effects are strongest in combating fatigue. With one exception, the evidence at this point does not support claims that it is possible to improve cognitive function in healthy adults on a day-to-day basis. Findings suggesting this might be the case often have not been replicated, many studies involve doses that may be unrealistic, and the effects observed in the lab may not translate to practical benefits. The exception is *Bacopa monnieri* where there are a number of studies with reasonably consistent findings of an improvement in memory, but the studies are relatively small and further research is needed to confirm these effects and examine whether they have an impact outside of the laboratory.

Effects observed are often stronger in those who are deficient in their diet, so it may be that it is not possible to improve cognitive function in people with a healthy, varied diet, and researchers need to measure dietary factors more systematically in intervention studies. Consistent with this suggestion, there is good epidemiological evidence that long-term dietary factors, including higher consumption than average of some specific foods, help to maintain cognitive function. There is relatively little evidence that pathological decline in ageing can be slowed by a dietary intervention once the condition has been diagnosed. Ginkgo biloba and cocoa, both sources of flavonoids, have the best evidence for such an effect, and flavonoids as a general class of substances are currently being investigated.

References

Atchison, T. B., Massman, B. J., & Doody, R. S. (2011). Baseline cognitive function predicts rate of decline in basic-care abilities of individuals with dementia of the Alzheimer's type. *Archives of Clinical Neuropsychology, 22*, 99–107. https://doi.org/10.1016/j.acn.2006.11.006

Bailey, R. L., Gahche, J. J., Miller, P. E., Thomas, P. R., & Dwyer, J. T. (2013). Why US adults use dietary supplements. *JAMA Internal Medicine, 173*, 355–361. https://doi.org/10.1001/jamainternmed.2013.2299

Bates, B., Lennox, A., Prentice, A., Bates, C., & Swan, G. (2011). National diet and nutrition survey. Headline results from Years 1, 2 and 3 (combined) of the rolling programme (2008/2009–2010/11). Retrieved from https://www.gov.uk/government/statistics/national-diet-and-nutrition-survey-headline-results-from-years-1-2-and-3-combined-of-the-rolling-programme-200809-201011

Bennett, T., & Holloway, K. (2017). Motives for illicit prescription drug use among university students: A systematic review and meta-analysis. *International Journal of Drug Policy, 44*, 12–22. https://doi.org/10.1016/j.drugpo.2017.02.012

Brownie, S. (2006). Predictors of dietary and health supplement use in older Australians. *Australian Journal of Advanced Nursing, 23*, 26–32.

Cacioppo, J. T., & Hawkley, L. C. (2009). Perceived social isolation and cognition. *Trends n Cognitive Sciences, 13*, 447–454. https://doi.org/10.1016/j.tics.2009.06.005

Cardello, A. V., & Schutz, H. G. (2003). The importance of taste and other product factors to consumer interest in nutraceutical products: Civilian and military comparisons. *Journal of Food Science, 68*, 1519–1524. https://doi.org/10.1111/j.1365-2621.2003.tb09677.x

Commenges, D., Scotet, V., Renaud, S., Jacqmin-Gadda, H., Barberger-Gateau, P., & Dartigues, J.-F. (2000). Intake of flavonoids and risk of dementia. *European Journal of Epidemiology, 16*, 357–363.

Cox, D. N., Koster, A., & Russell, C. G. (2004). Predicting intentions to consume functional foods and supplements to offset memory loss using an adaptation of protection motivation theory. *Appetite, 43*, 55–64. https://doi.org/10.1016/j.appet.2004.02.003

de Jong, N., Ocké, M. C., Branderhorst, H. A. C., & Friele, R. (2003). Demographic and lifestyle characteristics of functional food consumers and dietary supplement users. *British Journal of Nutrition, 89*, 273–281. https://doi.org/10.1079/BJN2002772

Denham, B. E. (2017). Psychosocial correlates of dietary supplement use: Results from the National Survey of Midlife Development in the United States. *Ecology of Food and Nutrition, 56*, 171–186. https://doi.org/10.1080/03670244.2016.1277711

Desideri, G., Kwik-Uribe, C., Grassi, D., Necozione, S., Ghiadoni, L., Mastroiacova, D., … Ferri, C. (2012). Benefits in cognitive function, blood pressure, and insulin resistance through cocoa flavanol consumption in elderly subjects with mild cognitive impairment: The Cocoa, Cognition, and Aging (CoCoA) study. *Hypertension, 60*, 794–801. https://doi.org/10.1161/HYPERTENSIONAHA.112.193060

Devore, E. E., Grodstein, F., van Rooij, F. J. A., Hofman, A., Stampfer, M. J., Witteman, J. C. M., & Breteler, M. M. B. (2010). Dietary antioxidants and long-term risk of dementia. *Archives of Neurology, 67*, 819–825. https://doi.org/10.1001/archneurol.2010.144

Devore, E. E., Kang, J. H., Breteler, M. M. B., & Grodstein, F. (2012). Dietary intakes of berries and flavonoids in relation to cognitive decline. *Annals of Neurology, 72*, 135–143. https://doi.org/10.1002/ana.23594

Dickinson, A., Blatman, J., El-Dash, N., & Franco, J. C. (2014). Consumer usage and reasons for using dietary supplements: Report of a series of surveys. *Journal of the American College of Nutrition, 33*, 176–182. https://doi.org/10.1080/07315724.2013.875423

DiNapoli, E. A., Wu, B., & Scogin, F. (2014). Social isolation and cognitive function in Appalachian older adults. *Research on Aging, 36*, 161–179. https://doi.org/10.1177/0164027512470704

Engelhart, M. J., Geerings, M. J., Ruitenberg, A., van Swieten, J. C., Hofman, A., Witteman, J. C. M., & Breteler, M. M. B. (2002). Dietary intake of antioxidants and risk of Alzheimer disease. *JAMA, 287*, 3223–3229.

Franke, A. G., Christmann, M., Bonertz, C., Fellgiebel, A., Huss, M., & Lieb, K. (2011). Use of coffee, caffeinated drinks and caffeine tablets for cognitive enhancement in pupils and students in Germany. *Pharmacopsychiatry, 44*, 331–338. https://doi.org/10.1055/s-0031-1286347

Helmer, S. M., Pischke, C. R., Van Hal, G., Vriesacker, B., Dempsey, R. C., Akvardar, Y., … Zeeb, H. (2016). Personal and perceived peer use and attitudes towards the use of nonmedical prescription stimulants to improve academic performance among university students in seven European countries.

Drug and Alcohol Dependence, 168, 128–134. https://doi.org/10.1016/j.drugalcdep.2016.08.639

Helmes, E., & Klinger, J. (2017). Prediction of everyday task performance in older adults by perceived health, self-efficacy and cognitive ability. *Cogent Psychology, 4*, 1297281. https://doi.org/10.1080/23311908.2017.1297281

Helvik, A.-S., Høgseth, L. D., Bergh, S., Šaltytė-Benth, J., Kirkevold, Ø., & Selbæk, G. (2015). A 36-month follow-up of decline in activities of daily living in individuals receiving domiciliary care. *BMC Geriatrics, 15*, 47. https://doi.org/10.1186/s12877-015-0047-7

Johnson, J. K., Lui, L.-Y., & Yaffe, K. (2007). Executive function, more than global cognition, predicts functional decline and mortality in elderly women. *The Journals of Gerontology, Series A: Biological Sciences and Medical Sciences, 62*, 1134–1141.

Kennedy, D. O., Scholey, A. B., & Wesnes, K. A. (2000). The dose-dependent cognitive effects of acute administration of Ginkgo biloba to healthy young volunteers. *Psychopharmacology, 151*, 416–423. https://doi.org/10.1007/s002130000501

Kennedy, D. O., Scholey, A. B., & Wesnes, K. A. (2002). Modulation of cognition and mood following administration of single doses of *Gingkgo biloba*, ginseng, and a ginkgo/Ginseng combination to healthy young adults. *Physiology & Behavior, 75*, 739–751. https://doi.org/10.1016/s0031-9384(02)00665-0

Kent, K., Charlton, K. E., Roodenrys, S., Batterham, M., Potter, J., Traynor, V., … Richards, R. (2017). Consumption of anthocyanin-rich cherry juice for 12 weeks improves memory and cognition in older adults with mild to moderate dementia. *European Journal of Nutrition, 56*, 333–341. https://doi.org/10.1007/s00394-015-1083-y

Kobayashi, E., Sato, Y., Umegaki, K., & Chiba, T. (2017). The prevalence of dietary supplement use among college students: A nationwide survey in Japan. *Nutrients, 9*, 1250. https://doi.org/10.3390/nu9111250

Kofoed, C. L. F., Christensen, J., Dragsted, L. O., Tjønneland, A., & Roswall, N. (2015). Determinants of dietary supplement use – Healthy individuals use dietary supplements. *British Journal of Nutrition, 113*, 1993–2000. https://doi.org/10.1017/S0007114515001440

Krikorian, R., Boespflug, E. L., Fleck, D. E., Stein, A. L., Wightman, J.-L. D., Shidler, M. D., & Sadat-Hossieny, S. (2012). Concord grape juice supplementation and neurocognitive function in human aging. *Journal of*

Agricultural and Food Chemistry, 60, 5736–5742. https://doi.org/10.1021/jf300277g

Krikorian, R., Nash, T. A., Shidler, M. D., Shukitt-Hale, B., & Joseph, J. A. (2010). Concord grape juice supplementation improves memory function in older adults with mild cognitive impairment. *British Journal of Nutrition, 103*, 730–734. https://doi.org/10.1017/S0007114509992364

Laurin, D., Masaki, K. H., Foley, D. J., White, L. R., & Launer, L. J. (2004). Midlife dietary intake of antioxidants and risk of late-life incident dementia: The Honolulu-Asia aging study. *American Journal of Epidemiology, 159*, 959–967. https://doi.org/10.1093/aje/kwh124

Laws, K. R., Sweetnam, H., & Kondel, T. K. (2012). Is *Ginkgo biloba* a cognitive enhancer in healthy individuals? A meta-analysis. *Human Psychopharmacology: Clinical and Experimental, 27*, 527–533. https://doi.org/10.1002/hup.2259

Letenneur, L., Proust-Lima, C., Le Gouge, A., Dartigues, J. F., & Barberger-Gateau, P. (2007). Flavonoid intake and cognitive decline over a 10-year period. *American Journal of Epidemiology, 165*, 1364–1371. https://doi.org/10.1093/aje/kwm036

Lieberman, H. R., Marriott, B. P., Williams, C., Judelson, D. A., Glickman, E. L., Geiselman, P. J., … Mahoney, C. R. (2015). Patterns of dietary supplement use among college students. *Clinical Nutrition, 34*, 976–985. https://doi.org/10.1016/j.clnu.2014.10.010

Loughrey, D. G., Lavecchia, S., Brennan, S., Lawlor, B. A., & Kelly, M. E. (2017). The impact of the Mediterranean diet on the cognitive functioning of healthy older adults: A systematic review and meta-analysis. *Advances in Nutrition, 8*, 571–586. https://doi.org/10.3945/an.117.015495

Mastroiacovo, D., Kwik-Uribe, C., Grassi, D., Necozione, S., Raffaele, A., Pistacchio, L., … Desideri, G. (2015). Cocoa flavanol consumption improves cognitive function, blood pressure control, and metabolic profile in elderly subjects: The Cocoa, Cognition, and Aging (CoCoA) study—A randomized controlled trial. *American Journal of Clinical Nutrition, 101*, 538–548. https://doi.org/10.3945/ajcn.114.092189

May, B. H., Lu, C., Lu, Y., Zhang, A. L., & Xue, C. C. L. (2013). Chinese herbs for memory disorders: A review and systematic analysis of classical herbal literature. *Journal of Acupuncture and Meridian Studies, 6*, 2–11. https://doi.org/10.1016/j.jams.2012.11.009

McCabe, S. E., Knight, J. R., Teter, C. J., & Wechsler, H. (2005). Non-medical use of prescription stimulants among US college students: Prevalence and correlates from a national survey. *Addiction, 99*, 96–106. https://doi.org/10.1111/j.1360-0443.2004.00944.x

Morris, M. A., Tangney, C. C., Wanga, Y., Sacks, F. M., Barnes, L. L., Bennett, D. A., & Aggarwal, N. T. (2015a). MIND diet slows cognitive decline with aging. *Alzheimer's & Dementia, 11*, 1015–1022. https://doi.org/10.1016/j.jalz.2015.04.011

Morris, M. A., Tangney, C. C., Wanga, Y., Sacks, F. M., Bennett, D. A., & Aggarwal, N. T. (2015b). MIND diet associated with reduced incidence of Alzheimer's disease. *Alzheimer's & Dementia, 11*, 1007–1014. https://doi.org/10.1016/j.jalz.2014.11.009

Neale, C., Camfield, D., Reay, J., Stough, C., & Scholey, A. (2012). Cognitive effects of two nutraceuticals Ginseng and Bacopa benchmarked against modafinil: A review and comparison of effect sizes. *British Journal of Clinical Pharmacology, 75*, 728–737. https://doi.org/10.1111/bcp.12002

Nehlig, A. (2010). Is caffeine a cognitive enhancer? *Journal of Alzheimer's Disease, 20*, S85–S94. https://doi.org/10.3233/JAD-2010-091315

Nooyens, A. C., Milder, I. E. J., van Gelder, B. M., Bueno-de-Mesquita, H. B., van Boxtel, M. P. J., & Verschuren, W. M. M. (2015). Diet and cognitive decline at middle age: The role of antioxidants. *British Journal of Nutrition, 113*, 1410–1417. https://doi.org/10.1017/S0007114515000720

Nurk, E., Refsum, H., Drevon, C. A., Tell, G. S., Nygaard, H. A., Engedal, K., & Smith, A. D. (2009). Intake of flavonoid-rich wine, tea, and chocolate by elderly men and women is associated with better cognitive test performance. *Journal of Nutrition, 139*, 120–127. https://doi.org/10.3945/jn.108.095182

O'Brien, S. K., Malacova, E., Sherriff, J. L., & Black, L. J. (2017). The prevalence and predictors of dietary supplement use in the Australian population. *Nutrients, 9*, 1154–1162. https://doi.org/10.3390/nu9101154

Pase, M. P., Kean, J., Sarris, J., Neale, C., Scholey, A. B., & Stough, C. (2012). The cognitive-enhancing effects of Bacopa monnieri: A systematic review of randomized, controlled human clinical trials. *The Journal of Alternative and Complementary Medicine, 18*, 647–652. https://doi.org/10.1089/acm.2011.0367

Reinert, A., Rohrmann, N., Becker, S., & Linseisen, J. (2007). Lifestyle and diet in people using dietary supplements: A German cohort study. *European Journal of Nutrition, 46*, 165–173. https://doi.org/10.1007/s00394-007-0650-2

Rendeiro, C., Rhodes, J. S., & Spencer, J. P. E. (2015). The mechanisms of action of flavonoids in the brain: Direct versus indirect effects. *Neurochemistry International, 89*, 126–139. https://doi.org/10.1016/j.neuint.2015.08.002

Riby, L. M. (2004). The impact of age and task domain on cognitive performance: A meta-analytic review of the glucose facilitation effect. *Brain Impairment, 5*, 145–165. https://doi.org/10.1375/brim.5.2.145.58253

Roodenrys, S., Booth, D., Bulzomi, S., Phipps, A., Micallef, C., & Smoker, J. (2002). Chronic effects of Brahmi (Bacopa monnieri) on human memory. *Neuropsychopharamacology, 27*, 279–281. https://doi.org/10.1016/S0893-133X(01)00419-5

Root, M., Ravine, E., & Harper, A. (2015). Flavonol Intake and Cognitive Decline in Middle-Aged Adults. *Journal of Medicinal Food, 18*, 1327–1332.

Sathyanarayanan, V., Thomas, T., Einöther, S. J. L., Dobriyal, R., Joshi, M. K., & Krishnamachari, S. (2013). Brahmi for the better? New findings challenging cognition and anti-anxiety effects of Brahmi (Bacopa monniera) in healthy adults. *Psychopharmacology, 227*, 299–306. https://doi.org/10.1007/s00213-013-2978-z

Singh, B., Parsaik, A. K., Mielke, M. M., Erwin, P. J., Knopman, D. S., Petersen, R. C., & Roberts, R. O. (2014). Association of Mediterranean diet with mild cognitive impairment and Alzheimer's disease: A systematic review and meta-analysis. *Journal of Alzheimers Disease, 39*, 271–282. https://doi.org/10.3233/JAD-130830

Singh, I., Bard, I., & Jackson, J. (2014). Robust resilience and substantial interest: A survey of pharmacological cognitive enhancement among university students in the UK and Ireland. *PLoS One, 9*, e105969. https://doi.org/10.1371/journal.pone.0105969

Smith, I., Williamson, E. M., Putnam, S., Farrimond, J., & Whalley, B. J. (2014). Effects and mechanisms of ginseng and ginsenosides on cognition. *Nutrition Reviews, 72*, 319–333. https://doi.org/10.1111/nure.12099

Sorond, F. A., Hurwitz, S., Salat, D. H., Greve, D. N., & Fisher, N. D. L. (2013). Neurovascular coupling, cerebral white matter integrity, and response to cocoa in older people. *Neurology, 81*, 904–909. https://doi.org/10.1212/WNL.0b013e3182a351aa

Stuck, A. E., Walthert, J. M., Nikolaus, T., Büla, C. J., Hohmann, C., & Beck, J. C. (1999). Risk factors for functional status decline in community-living elderly people: A systematic literature review. *Social Science & Medicine, 48*, 445–469.

Teter, C. J., McCabe, S. E., LaGrange, K., Cranford, J. A., & Boyd, C. J. (2006). Illicit use of specific prescription stimulants among college students:

Prevalence, motives, and routes of administration. *Pharmacotherapy, 26*, 1501–1510.

Vauzour, D. (2014). Effect of flavonoids on learning, memory and neurocognitive performance: relevance and potential implications for Alzheimer's disease pathophysiology. *Journal of the Science of Food and Agriculture, 94*, 1042–1056. https://doi.org/10.1002/jsfa.6473

Wang, X., Ouyang, Y. Y., Liu, J., & Zhao, G. (2014). Flavonoid intake and risk of CVD: A systematic review and meta-analysis of prospective cohort studies. *British Journal of Nutrition, 111*, 1–11. https://doi.org/10.1017/S000711451300278X

Zahodne, L. B., Manly, J. J., MacKay-Brandt, A., & Stern, Y. (2013). Cognitive declines precede and predict functional declines in aging and Alzheimer's disease. *PLoS One, 8*(9), e73645. https://doi.org/10.1371/journal.pone.0073645

5

Being Limitless: A Discursive Analysis of Online Accounts of Modafinil Use

Matthew Hall, Mark Forshaw, and Catharine Montgomery

Introduction

The war on Alzheimer's and other cognitive progressive neurological diseases which affect multiple brain functions, such as memory, has stimulated an intensive effort to develop drugs that improve cognitive function—known as cognitive enhancement (Dubljević, Venero, & Knafo, 2015). According to Förstl (2009) there are now more than 100 cognitive enhancement drugs either being developed, or tested, or in use.

M. Hall (✉)
Arden University, Coventry, UK

University of Derby, Derby, UK
e-mail: mhall@arden.ac.uk

M. Forshaw • C. Montgomery
Liverpool John Moores University, Liverpool, UK
e-mail: M.J.Forshaw@ljmu.ac.uk; C.A.Montgomery@ljmu.ac.uk

© The Author(s) 2021
M. Hall et al. (eds.), *Chemically Modified Minds*,
https://doi.org/10.1007/978-981-15-6771-1_5

Some of those covered in this book include Noopept;[1] Phenylpiracetam;[2] Pramiracetam;[3] Aniracetam;[4] and Piracetam.[5] One of the more popular compounds is modafinil, the focus of our chapter.

Modafinil (trade name Provigil) is a eugeroic agent that increases levels of cortical catecholamine, serotonin, glutamate, orexin, and histamines whilst decreasing the level of gamma-aminobutyric acid (Battleday & Brem, 2015). These changes stimulate the central nervous system with the user experiencing increased wakefulness, alertness, and cognitive enhancement. Hou et al. (2005) study of healthy male volunteers found that the maximum safe dosage of 200 mg of modafinil per day administered in a single tablet in the morning resulted in stimulation of the locus coeruleus a wakefulness-promoting noradrenergic nucleus.

Given fatigue poses a serious threat to operational safety in some employments, studies have explored the use of modafinil by military personnel, medical professionals (Westcott, 2005), and pilots (Caldwell, Caldwell, Smythe, & Hall, 2000; Yesavage et al., 2002). Controlled investigations that tested performance at intervals over 37–48-hour periods found that cognitive performance for those that were administered a daily dose of modafinil increased by 15–30% compared to those who took a placebo that showed 60–70% decline. However, several studies have shown that although stimulating neurotransmitter systems can enhance cognition, mood, and pro-social behaviour, an increased performance in one cognitive domain was often found to coincide with a decrease in performance in another domain (de Jongh, Bolt, Schermer, &

[1] Noopept is a peptide-derived nootropic that is closely related to the racetam family carrying a similar method of action but with a higher bioavailability.

[2] Phenylpiracetam is a nootropic of the racetam family synthesized from piracetam. Phenylpiracetam is reported to be more neuroprotective than piracetam, enhancing physical performance and also possessing psychostimulatory properties.

[3] Pramiracetam is a nootropic of the racetam family synthesized from piracetam. Pramiracetam is reported to have anti-amnesiac potential, improved long-term memory, and cognitive performance. However, most studies have involved rats.

[4] Aniracetam is a racetamic and Ampakine drug that has a marginally higher in potency than Piracetam. The substance has similar effects to Noopept; Phenylpiracetam; Pramiracetam and; Piracetam but with the addition of observed anxiety-reducing effects.

[5] Piracetam is a nootropic of the racetam family. Piracetam is reported to improve long-term memory and cognitive performance.

Olivier, 2008, p. 762). For example, working memory may be enhanced at the cost of ong-term memory. de Jongh et al. (2008) also point out that whilst people with a low memory span might benefit, those high span memory subjects may be impaired through overdosing.

There have been increasing concerns about other negative effects of the modafinil too. This has resulted in changes in medical use. Previously modafinil was prescribed in the UK for somnolence experienced by shift-workers and for those with obstructive sleep apnoea. However, in March 2011 the UK Medicines and Healthcare Products Regulatory Agency restricted the use of modafinil to narcolepsy following a European Medicines Agency (2010) review of the safety and effectiveness of modafinil. The Committee for Medicinal Products for Human Uses (CHMP) Pharmacovigilance Working Party (2010, p. 1) found modafinil was:

> strongly linked to a risk of serious, life-threatening skin reactions, and this risk appears higher in children. The Committee also noted a link between modafinil and psychiatric adverse reactions, such as suicidal thoughts, depression, psychotic episodes, and between modafinil and cardiovascular adverse reactions, such as hypertension (high blood pressure) and irregular heartbeat.

Serious psychiatric disorders include suicidal thoughts, mania, and symptoms of psychosis such as delusion, and skin reactions include allergic reactions which may be severe such as Stevens-Johnson syndrome, which affects the skin and mucous membranes. Thus, the CHMP concluded that for use other than narcolepsy the risks outweighed the benefits.

But despite these risks there is little data available on the long-term effects of their use is or how widespread their use is. Chatterjee (2007, 2013) suggests the use of cognitive enhancers is likely to increase, similarly to cosmetic surgery, as bioethical and psychological concerns are overcome with cultural acceptance. Indeed, Dubljević et al. (2015) suggest that it is likely that 5–15% of the US students have used cognitive enhancers at some point. A poll conducted by the journal *Nature* (Maher, 2008, p. 1) of 1400 readers from 60 countries on the use of three well-known performance-enhancing agents—methylphenidate (Ritalin[6]), modafinil, and beta blockers[7]—found one in five respondents said they had "used drugs for non-medical reasons to stimulate their focus, concentration or memory". No age-related differences were found. The most popular reason for taking substances was to improve concentration. When asked about their attitudes towards neuroenhancing substances, four-fifths thought that usage was down to the individual and 69% would risk mild side effects to take such substance. Given the tension between health risk and enhanced cognitive performance, how people talk about modafinil use becomes an important question.

Method

Data

In the UK, modafinil is a prescription-only product. However, modafinil is widely available online without prescription as a "smart drug" to promote cognitive enhancement. Several Internet sites such as MedPillRx.com and the Online Pharmacy (http://modafinil24h.com/), often located in the US and Canada, offer modafinil for as little as $0.88 per pill. Consumers can purchase from 10–380 pills at strengths of 100 mg, 150 mg, or 200 mg. Internet sites such as Modafinil Rocket provide users with information on country-specific legislation, links to websites selling modafinil, product reviews, and tips and tricks for use.

Our data is drawn from the Modafinil Forum,[8] an electronic bulletin board where members can begin threads for the purpose of discussion, building bonds, and reaching others interested in the topic of modafinil

---

[6] Ritalin is a central nervous system stimulant that contributes to hyperactivity and impulse control. Thus, as a performance enhancer, the person maintains a state of calm as well as increased brain activity.

[7] Beta-blockers such as Propranolol treat tremors, angina, hypertension, heart rhythm disorders, and other heart or circulatory conditions. It is used as a performance enhancer in reducing the symptoms of stress by maintaining a regular heart rhythm.

[8] Since the beginning of 2018, Modafinil Forum has closed.

use. We collected data (September 23, 2017) from the most popular thread "Being Limitless" containing 52 posts by 32 respondents between September 30, 2014, and May 14, 2017.

Ethics

Collecting data from the Internet presents ethical challenges around what is deemed a "public" or "private" space. One obvious issue is whether informed consent can be gained. Some scholars (e.g. Hookway, 2008) argue that open access online discussion boards, forums, and blogs are firmly located in the public domain. As such contributors are aware that their posts will be read by others unless they place them on a "friends only" setting. Thus, accessible electronic talk may be 'personal' but it is not 'private' (Hookway, 2008: 16) and so consent can be 'waived'. However, given the potentially sensitive nature of our data we deemed it appropriate to seek university ethical approval, and in line with BPS guidelines (BPS, 2017), we have anonymised our dataset as far as possible (e.g. replacing tags and pseudonyms with R1 [Respondent 1], R2, etc., removing any in-text personal details or references).

Data Analysis

Having downloaded all posts, we then examined the data-set using discourse analysis as set out by Jonathan Potter (1996). Broadly speaking, discourse analysts aim to explore how 'versions of world, of society, events and inner psychological worlds are produced in discourse' (Potter, 1996, p. 146). People work up versions of the world during online (and offline) interactions. The particular version will be dependent on the topic of conversation (e.g. motivation for use), who the other interactant(s) are (e.g. fellow modafinil users), the context (e.g. modafinil experiences), location (e.g. modafinil specific forum), and time (e.g. recent trend). These versions of the world can be seen in talk-in-interaction by following (Baker, 1997) three-step analytical process: locate the central themes that are named and/or implied in the talk; focus on the discursive activities within each segment of talk; and examine how the interactants construct their accounts, produce descriptions of events and activities, manage their personal interest in the event or activity, and how these are framed and for what purpose(s). However, as Edwards and Potter (1992) point out, during the analytical process, analysts should only read what is made relevant by the interactants to avoid analyst-lead interpretations.

Results

Posters responded to the forum question which asked users why they take modafinil (e.g. a student, business owner, work long hours) and whether they take it all of the time or at certain times (e.g. for a business deal, or exam). In analysing posters responses, we noted that this was a 'community of practice' (e.g. Ba, 2001) where forum contributors worked up their community membership. Respondents talked about using modafinil for ambition, to help with technical work, concentration, energy, focus, social proficiency, to suppress the effects of alcohol, increase passion, commitment, and drive, aid efficiency, creativity, parenting, and to enhancing well-being. We explore how respondents work up accounts of their modafinil use as credible, authentic, and legitimate in the following analysis (Epstein, 1995). Given book chapter word limitations, we selected posts and sequences to show the various motivations for using this substance.

We begin our analysis with a sequence in which two self-identified modafinil users discuss and legitimise their use of this substance to challenging social barriers and aiding a demanding work schedule:

Ambitions, and Technical Work

R1: I am an engineering student with ambitions to pursue graduate studies at either Caltech or MIT. Ultimately it is not really about what school I go to, but the information I attain from education. Learning will help me with bigger goals, like starting a car company or inventing new technology. Innovation takes a toll on the brain; makes you want to fall asleep. That is where Modafinil comes in

R2: I am a practicing engineer working on some very demanding projects at the moment requiring long hours of technical work that really take a mental toll, this is where Modafinil come in and it really help with this kind of work.

Several things are immediately visible in R1's post. R1 wants to pursue education "graduate studies" at either "Caltech or MIT"—top-tier US universities. However, anticipating that other readers might question the plausibility of this ambition, R1 works up an account in which access to them centres on cognitive ability "the information I attain from education" rather than R1's current educational establishment "it is not really about what school I go to". This is bolstered by the deployment of the extreme-case formulation 'really' (Pomerantz, 1986). As Edwards (2000, p. 348) points out, extreme-case formulations such as 'never', 'always', 'none', and 'all' serve as discursive devices in 'defending positions against refutation, making complaints, and justifying factual claims'. R1's use of "really" can be seen as defending his position against potential refutation, whilst also working up a position in which using modafinil as a means to access top-tier universities is a legitimate reason.

What is also noticeable is that R1 further links modafinil use to other successes such as achieving "bigger goals" such as "starting a car company, or inventing new technology". In doing so, R1 works up a position in which modafinil use challenges social barriers such as where one goes to "school" by enhancing one's cognitive ability. Substance use downsides are not discussed and indeed, mental fatigue from "Innovation" is problematized "takes a toll on the brain; makes you want to fall asleep". However, readers of R1's post might question its credibility as it does not provide evidence of success; only what is presumed will happen.

R2's immediate response adds credibility to R1's claim, "Innovation takes a toll on the brain" by providing experiential evidence. Epstein (1995) notes that one of the defining features of credibility is the presentation of factual knowledge. That is, R2 is a "practicing engineer" rather than an "engineering student" and thus claims to have first-hand experience "I am…working on some very demanding projects at the moment requiring long hours of technical work that really take a mental toll, this is where Modafinil come in". By drawing on his own experience and knowledge, R2 provides support and so constructs a second story (Veen, te Molder, Gremmen, & van Woerkum, 2010). That is, the alignment of a second response to the original response (Sacks, 1992). Second stories work to normalise views by displaying an understanding and stance towards the initial story and they also work to add credibility by drawing experiential knowledge (Arminen, 2004). Therefore, by drawing on experiential knowledge, R2 lends credibility to the claims about the benefits of modafinil use.

R3's response below also discusses the cognitive impact of modafinil use, but whereas R1 and R2 highlight modafinil's ability to combat mental fatigue, R3 focuses on the drug's capacity to aid concentration and provide energy.

Concentration, and Energy

R3: I use it to stay awake. I don't get a full 8 hours of sleep sometimes not even 6 and Modafinil helps me from falling asleep as well as concentrating. I might have a problem with falling asleep as well. It's done wonders and I can work effortlessly without feeling tired. I mainly use it for work + running a side business and having the energy to complete tasks which I personally don't enjoy.

R3's post opens with a statement of use 'I use it to stay awake', which highlights one aspect of modafinil's impact on the body 'stay awake' (see Hou et al. 2005) study of healthy male volunteers for more on modafinil and wakefulness). When people make statements, they are compelled to provide a legitimate account for doing so (Potter, 1996). R3 proceeds to provide an account for needing to stay awake, which centres on a problem "I don't get a full 8 hours of sleep sometimes not even 6" and as a result "falling asleep". Although R3 highlights the potential downside of consuming a psychoactive stimulant "I might have a problem with falling asleep as well", R3 immediately proceeds to present a list of presumed beneficial impacts "stay awake", "concentrating", "work effortlessly", and "energy to complete tasks". Jefferson's (1991, p. 68) work on list construction showed that they can be used as an 'orientated-to-procedure'. In other words, it provides the others with a means to discursively position themselves in relation to the items on the list, such that they can either ascribe to or disavow membership, based on the items provided. Moreover, the list provides other forum members, whether modafinil users or not, to accord group status. That is, whether their modafinil use is legitimate. Jefferson (1991) also noted that lists serve to normalise the cited practices, thereby attempting to remove uncertainty. However, as Jefferson also noted, a list is always contestable, because the items can be viewed as inappropriate or not legitimate. Thus, R3 provides an account for seeking these benefits "I mainly use it for work + running a side business and having the energy to complete tasks which I personally don't enjoy". In other words, the effects of modafinil help to combat the downsides of a busy lifestyle.

Similarly, to R3, R4's post below is also constructed around modafinil use for a busy lifestyle but with the added benefit of improving interactions with clients:

Focus, and Social Proficiency

R4: I run a business. Typically, without moda I am falling asleep after lunch, with moda I have no such problems. It helps me stay awake and focused all day. I find it also helps when dealing with clients, I feel more sharp and on point socially, which makes it easier to close sales.

R4's opening statement "I run a business" provides readers with a reference point in which to assess R4's modafinil use. It also situates running a business as a normative activity that invokes category-bound activities and predicates (see Sacks, 1992 for more on membership categorisation) such as being busy, interacting with staff and clients, management, multitasking, selling, and so on—all typical markers of running a business (see Drucker, 1986). In doing so, R4 is inviting readers to associate "falling asleep after lunch" as a symptom of running a business without explicitly saying so. Sacks (1992) suggests that we hear it this way because of the 'hearer's maxim'. That is, people hear collection of categories as going together (e.g. manager, clients) and that there are normative relationships between identities (e.g. someone who runs a business) and the category-bound activities and predicates associated with that (e.g. being busy, dealing with clients) without them having to be said out aloud. Thus, tiredness and needing to "stay awake and focused all day" are presented as legitimate reasons for R4's modafinil use. However, "stay awake and focused all day" are also items on R4's three-part list of positive reasons for using modafinil which also includes "helps when dealing with clients". Jefferson (1991) showed, the presence of three items, or more, on a list adds clarity and weight to arguments. In other words, strength by numbers. But, despite these positive aspects, in order to be considered legitimate and credible, some readers might require clarity on how wakefulness and focus translate into better interactions with clients and ultimately "easier to close sales". Thus, R4 highlights the presumed positive psychosocial effects of modafinil "I feel more sharp and on point socially" that can be usefully deployed to a work context to improve productivity.

In the following post, the poster works up an account in which the effects of modafinil help to stay focused 'without getting too distracted' and avoid the 'drawbacks of alcohol', hearable as drunkenness (Sacks, 1992):

Focus, and Alcohol Suppression

R5: I use it for work, I work at a university lab while doing a PhD and take it 3 times a week. It allows me to get through the day without getting too distracted, as I love to side track in my research and can sometimes take a whole day trying to do something completely unrelated to my topic if I don't take it. I also take it before a night out as it allows me to drink without experiencing drawbacks of alcohol.

Like other users' accounts, modafinil use in R5's account is presented as having positive impacts that remedy problems such as "getting too distracted" and "experiencing drawbacks of alcohol". But what is interesting about R5's post is the way work "I work at a university lab" and education "doing a PhD" index his intellectual credentials and thus, his authority, expertise, and experience in modafinil use. R5's credibility is also indexed in other ways. For example, by highlighting knowledge of use 'I…take it 3 times a week', and highlighting situated expertise (Mackiewicz, 2010), in an employment 'I use it for work' and social context 'I…take it before a night out', both help to provide authenticity. Although R5's account is on a personal 'footing' (Goffman, 1979) 'I', 'me', and 'my', in a 'community of practice' (e.g. Ba, 2001) context where other forum members are of self-identified 'modafinil users' R5's account, like other contributors, may be viewed as presenting modafinil as a 'safe to use' substance.

So far only one potential downside of this psychoactive stimulant has been highlighted "I might have a problem with falling asleep" (R3). In R6's post below we also get a sense that the elevated focus and wakefulness may also negatively impact on tolerance levels "I don't have a lot of patience for people's nonsense".

Passion, Commitment, and Drive

R6: I use Modafinil for everything I do in my busy day. It gives me an intense focus, almost like a state of hypnosis. At the gym I am super focused on my intense muscular contractions of my HIT workouts. No wasted energy at all, it locks me into my music and getting down to kicking ass! At work it is as if I can really put the pedal to the metal n grind out long days and come home and do research and writing with mental clarity and drive. I noticed that I don't have a lot of patience for people's nonsense because I want to get things done without any interference. That can be challenging when you have to deal with people who move and think like snails. It allows me to take on new challenges with passion and commitment. I'm an intense person and it gives me heightened sense of being in a groove with a take no prisoners attitude.

R6 presents a list of contexts in which modafinil is presumed to be beneficial, "At the gym", "At work", to "do research" and "writing", as well as a list of positive ways that modafinil helps "like a state of hypnosis", "No wasted energy", "pedal to the metal", "mental clarity and drive", "passion and commitment", "heightened sense of being in a groove", and a "take no prisoners attitude". Listing, as we have previously noted, does several things: it helps to bolster an account through strength by numbers; it provides readers with an 'orientated-to-procedure' that they can position themselves in relation to the items on the list (e.g. (dis)avow them) (Jefferson, 1991, p. 68) and; it evokes truth(s) based on (R6's) experiential knowledge (Lindstrom, 1992).

But, what is also noticeable is that R6 sets up a contrast pair (Smith, 1978)—modafinil users ('mental clarity and drive', 'passion and commitment')/non-modafinil users ('people who move and think like snails'). The activity serves to hold non-users accountable for not having the same sense of drive, passion, and commitment and thus having inferior self-respect—people R6 does not "have a lot of patience for". Although R6's contrastive pair is explicitly invoked, we can see similar contrastive pairs in all hitherto posts, such as ambitious/less ambitious (R1), knowledgeable/less informed (R1, R2, R5), busy/less active (R1–5), and more sociable/less sociable (R4, R5). In doing so forum contributors are working up accounts in which they set themselves apart from others.

In the following post, R7 works up contrast pair relating to the quality of parenting happy mums/unhappy mums:

Efficiency, Creativity, and Parenting

R7: I use mod because it makes me the bestest, funnest, happiest Mom in the world! Lmao! In all seriousness, on the days I take mod, I am much more efficient, in a great mood, and I feel that I am more creative too. We crank out some awesome arts and crafts on those days! On those days, somehow the majority of chores get done, the kids get more attention, and overall ... simply, I am happy.

R7's post opens with humour, indexed with 'Lmao!' (a colloquial acronym for 'laugh my ass off') and bolstered with a three-part list of attributes: 'bestest, funnest, happiest' (Jefferson, 1991). Benwell's (2004) study of humour and irony shows that people use it to reduce the risk of being taken too seriously, and is often deployed in contexts when delicate, sensitive, or taboo topics are being discussed. In this post R7 self-identifies as a "Mom" who takes "mod". Hogan's (2003) work on parents who consume psychoactive substances shows that they are generally viewed as bad parents, and more so for mothers than fathers. Thus, the immediate deployment of humour works to avoid R7 being re-categorised as a bad mother (Speer, 2005).

Having started the post on a humorous note, R7 is then in a position to present the positive aspects of modafinil whilst reducing the potential for others to criticise. The benefits of modafinil are bolstered with a three-part list of effects 'I am much more efficient, in a great mood, and I feel that I am more creative too' (Jefferson, 1991). But, being aware of the delicacy of talking seriously "In all seriousness" about modafinil as a parent, R7 downplays her use as minimal "on the days I take mod" whilst accentuating the benefits for her children "kids" such as "We crank out some awesome arts and crafts", "chores get done", "kids get more attention", and "I am happy". In other words, R7 is able to work up an account of her substance use as unproblematic framing it as beneficial for her children, and in doing so positioning herself as a good parent.

Feeling upbeat "happy" seemed to underlie many the majority of posts. In the following post, the respondent links modafinil use with ageing and vitality "I'm in my 50s and have never felt as alive and well as I have for the last 3 years while taking modafinil":

Age and Vitality

R8: I take modafinil for everyday life! I'm in my 50s and have never felt as alive and well as I have for the last 3 years while taking modafinil. Modafinil boosts my energy and ability to focus on everyday tasks at work and at home. I know there are other people like me who don't have very demanding lifestyles but who love the focus they get from modafinil. I would love to hear from these people about their experiences and the difference modafinil has made in their lives.

What is immediately noticeable about R8's post is that modafinil usage is not aligned with a busy lifestyle like other posters "I know there are other people like me who don't have very demanding lifestyles". Thus, although R8 is a member of the category "modafinil users", R8's experience is different in this respect. That is, R8 is not taking this substance as a means to help manage multiple tasks and activities such as parenthood, work, study, socialising, and fitness.

We get a sense that lifestyle is related to age because R8 makes relevant age group "I'm in my 50s". In developed nations ageing is often associated with a slower lifestyle, which may be accompanied by physical and cognitive decline (Calasanti, 2005). In making age relevant, R8 is able to work up an account in which modafinil helps to contest the ageing process "Modafinil boosts my energy and ability to focus on everyday tasks at work and at home". R8's account is bolstered with the longevity of use "3 years while taking modafinil". Thus, R8's experiential knowledge adds credibility to the claim whilst also implying that the substance is safe to use by providing experiential evidence (Epstein, 1995). What is also interesting is that by invoking this age group, R8 implies that consumption of this pharmaceutical is likely to be by younger generations. Although little is known about the use of this substance by older generations, Rao and Crome (2017, p. 358) point out that older generations are increasingly vulnerable to prescription drug misuse and estimates suggest that 'the number of people aged over 50 using substances problematically is increasing across a range of settings globally'.

Discussion

In this chapter we have focused on modafinil users' motivations and how this community of users worked up accounts of this prescription-only substance as credible, authentic, and legitimate: a community of practice. Our analysis of posts showed the majority of users did so for rewards such as helping focus and concentration on technical work and study, or providing the user with energy and a sense of well-being whilst managing multiple facets of their, often busy, lifestyle. Indeed, for some, this pharmaceutical was used to counteract the unwanted side effects of other substances such as alcohol. Thus, the commonality between these accounts was that modafinil use had positive impacts across a range of areas of their lives.

Unsurprisingly, given the context of the forum, there was a relative absence of any reported side effects especially given some consumers reported using this substance regularly for a number of years. But what was also had a relative absence was discussion on how to take this substance. For example, appropriate dosages, when to take, length of cycle—frequency, longevity of the substance in the body, the anatomical effects, and its comparison to other pharmaceuticals. This surprised us, especially since some users had been doing so for a number of years. Other studies of substance misuse (e.g. Hall, Grogan, & Gough, 2016) show that this type of experiential information is willingly shared in communities of users so other members can reduce risks associated with consumption, for example overdosing. This is especially prevalent where publicly available medical and pharmaceutical information on the substance is not ubiquitous or clear. This suggests to us that this cognitive enhancing substance seems to be viewed as relatively safe to consume (e.g. see poll conducted by the journal *Nature*; Maher, 2008). But, given the European Medicines Agency (2010) review of the safety and effectiveness of modafinil, and the substances links to serious, life-threatening side effects such as skin reactions, suicidal thoughts, depression, psychotic episodes, and cardiovascular problems, we think more needs to be done to challenge these views, especially since modafinil is widely available on the Internet.

Thus, from a health promotion perspective, information on the dangers of modafinil which cites scientific evidence could be posted online or in spaces where usage is presumed to be common such as at universities. This could include posters for use in citing scientific evidence for dangers of use, ideally supported by endorsements from those likely to be perceived to be "experts" and who are trusted, such as successful business people and academics. Similar materials could raise questions about the ethical and moral dimensions of using substances to enhance cognitive performance (Chatterjee, 2013). As well, alternative nonchemical ways of enhancing cognitive performance could be promoted at these sites, supported by those who are seen as having credibility within specific communities. Piloting these materials ahead of use with those who have used this substance will be crucial in ensuring that messages are perceived as relevant and credible by those in the target group. This work also has important implications for promoting health in users of other "smart drugs". Clearly there is a need for medically accurate information that has credibility with such users to enable users to have somewhere to go to find accurate information on health risks. This information needs to recognise reasons for use, and be represented using language that has credibility among this community.

Although our study works with original, naturalistic data around a relatively new but poorly understood phenomenon, we recognise that our study is preliminary and that much more research is required with "smart drug" users. For example, although we suspect, based on our data set, that most users are younger generations, for example pre-40-year-olds, we cannot be sure of this given the methodology used here, and so gaining access to users in other contexts for surveys and interviews would help us to determine this demographic, and to gain the perspectives of others from different age (and ethnic, class) groups concerning perceived risks and benefits. In contrast to public posts online, one-to-one interviews would allow people time and space to account for their "smart drug" use in detail and in confidence. Finally, it would also be interesting to conduct qualitative longitudinal case studies with users whereby patterns of use and associated effects on performance and cognition could be tracked over time. Such research is likely to be timely, and urgent, given lifestyles are reported to be busier and cognitive performance is associated with life outcomes such as income levels (Chatterjee, 2007, 2013).

References

Arminen, I. (2004). Second stories. The saliency of interpersonal communication for mutual help in alcoholics anonymous. *Journal of Pragmatics, 36*, 319–347.

Ba, S. (2001). Establishing online trust through a community responsibility system. *Decision Support Systems, 31*(3), 323–336.

Baker, C. (1997). Membership Categorization and Interview Accounts. In D. Silverman (Ed.) *Qualitative Research: Theory, Method and Practice*. London: Sage.

Battleday, R. M., & Brem, A. K. (2015). Modafinil for cognitive neuroenhancement in healthy non-sleep-deprived subjects: A systematic review. *European Neuropsychopharmacology, 26*(2), 1865–1881.

Benwell, B. (2004). Ironic discourse: Evasive masculinity in British men's lifestyle magazines. *Men and Masculinities, 7*(1), 3–21.

British Psychological Society (2017). *Guidelines for Internet-mediated Research*. Retrieved April 21, 2019 from https://www.bps.org.uk/sites/www.bps.org.uk/files/Policy/Policy%20-%20Files/Ethics%20Guidelines%20for%20Internetmediated%20Research%20%282017%29.pdf

Calasanti, T. (2005). Ageism, gravity, and gender: Experiences of aging bodies. *Generations, 29*(3), 8–12.

Caldwell, J. A., Caldwell, J. L., Smythe, N. K., & Hall, K. K. (2000). A double-blind, placebo-controlled investigation of the efficacy of modafinil for sustaining the alertness and performance of aviators: A helicopter simulator study. *Psychopharmacology, 150*(3), 272–282.

Chatterjee, A. (2007). Cosmetic neurology and cosmetic surgery: Parallels, predictions, and challenges. *Cambridge Quarterly of Healthcare Ethics, 16*(2), 129–137.

Chatterjee, A. (2013). The ethics of neuroenhancement. *Handbook of Clinical Neurology, 118*, 323–334.

de Jongh, R., Bolt, I., Schermer, M., & Olivier, B. (2008). Botox for the brain: Enhancement of cognition, mood and pro-social behavior and blunting of unwanted memories. *Neuroscience & Biobehavioral Reviews, 32*(4), 760–776.

Drucker, P. F. (1986). *Management: Tasks, responsibilities, practices*. New York: Truman Talley Books / E.P. Dutton.

Dubljević, V., Venero, C., & Knafo, S. (2015). What is cognitive enhancement? In S. Knafo & C. Venero (Eds.), *Cognitive enhancement: Pharmacologic, environmental and genetic factors* (pp. 1–9). Cambridge, MA: Academic Press.

Edwards, D. (2000). Extreme Case Formulations: Softeners, Investment, and Doing Nonliteral. *Research on Language & Social Interaction, 33* (4), 347–373.

Edwards, D., & Potter, J. (1992). *Discursive psychology: Inquiries in social construction*. London: Sage.

Epstein, S. (1995). Expertise: AIDS activism and the forging of credibility in the reform of clinical trials. *Science, Technology & Human Values, 20*(4), 408–437.

European Medicines Agency. (2010). *Modafinil*. Retrieved October 7, 2017 from http://www.ema.europa.eu/ema/index.jsp?curl=pages/medicines/human/referrals/Modafinil/human_referral_000236.jsp

European Medicines Agency. (2010). *Committee for Medicinal Products for Human Use (CHMP) 15 – 19 November 2010*. Retrieved April 21, 2019 from https://www.ema.europa.eu/en/documents/committee-report/committee-medicinal-products-human-use-chmp-november-2010-plenary-meeting-monthlyreport_en.pdf

Förstl, H. (2009). Neuro-enhancement: Brain doping. *Nervenarzt, 80*(7), 840–846.

Goffman, E. (1979). Footing. *Semiotica, 25*, 1–29.

Hall, M., Grogan, S., & Gough, B. (eds.) (2016). *Chemically modified bodies: The use of diverse substances for appearance enhancement*. London: Palgrave Macmillan.

Hogan, D. M. (2003). Parenting beliefs and practices of opiate-addicted parents: Concealment and taboo. *European Addiction Research, 9*, 113–119.

Hookway, N. (2008). Entering the blogosphere. Some strategies for using blogs in social research. *Qualitative Research, 8*(1), 91.

Hou, R. H., Freeman, C., Langley, R. W., Szabadi, E., & Bradshaw, C. M. (2005). Does modafinil activate the locus coeruleus in man? Comparison of modafinil and clonidine on arousal and autonomic functions in human volunteers. *Psychopharmacology, 181*(3), 537–549.

Jefferson, G. (1991). List construction as a task and a resource. In G. Psathas (Ed.) *Interactional Competence*. New York: Irvington Publications.

Lindstrom, L. (1992). Context contests: Debateable truth statements on Tanna (Vanuata). In A. Duranti (Ed.), *Rethinking context: Language as an interactive phenomenon (Studies in the social and cultural foundations of language)* (pp. 101–124). Cambridge: Cambridge University Press.

Mackiewicz, J. (2010). The Co-construction of credibility in online product reviews. *Technical Communication Quarterly, 19*(4), 403–426.

Maher, B. (2008). Poll results: Look who's doping: In January, nature launched an informal survey into readers' use of cognition-enhancing drugs. Brendan Maher has waded through the results and found large-scale use and a mix of attitudes towards the drugs. *Nature, 452*(7188), 674–675.

Pomerantz, A. (1986). Extreme case formulations: a way of legitimizing claims. *Human Studies, 9*, 219–229.

Potter, J. (1996). *Representing reality: Discourse, rhetoric and social construction*. London: Sage.

Rao, R., & Crome, I. (2017). Substance misuse in older people. *British Medical Journal*. Retrieved from https://www.bmj.com/content/358/bmj.j3885

Sacks, H. (1992). *Lectures on conversation*. Ed. G. Jefferson. Oxford: Blackwell Publishing.

Smith, D. (1978). K is mentally ill: The anatomy of a factual account. *Sociology, 12*, 23–53.

Speer, S. (2005). *Gender talk: Feminism, discourse and conversation analysis*. Hove: Routledge.

Veen, M., te Molder, H., Gremmen, B., & van Woerkum, C. (2010). Quitting is not an option. An analysis of online diet talk between celiac disease patients. *Health, 14*(1), 23–40.

Westcott, K. J. (2005). Modafinil, sleep deprivation, and cognitive function in military and medical settings. *Military Medicine, 170*(4), 333–335.

Yesavage, J. A., Mumenthaler, M. S., Taylor, J. L., Friedman, L., O'Hara, R., Sheikh, J., … Whitehouse, P. J. (2002). Donepezil and flight simulator performance: Effects on retention of complex skills. *Neurology, 3*(1), 123–125.

6

The Role of Glycaemic Control in Cognitive Functioning

Jade M. Elliott

Introduction: What Is Glycaemic Control and Why Is Glucose Critical to Cognitive Functioning?

The brain is the most metabolically expensive organ within the body, requiring constant energy provision, even at times of rest. Despite accounting for a mere 2% of body weight, the brain demands approximately 20% of the basal metabolic rate (Benton, 2001). The main energy substrate for the brain is glucose, which is a simple sugar or monosaccharide. Although in times of extreme starvation, the brain can use other substrates of lactate, pyruvate, and ketone bodies (Bélanger, Allaman, & Magistretti, 2011; Languren, Montiel, Julio-Amilpas, & Massieu, 2013; Magistretti, 2008).

J. M. Elliott (✉)
Department of Psychology, Staffordshire University, Staffordshire, UK
e-mail: J.elliott@staffs.ac.uk

© The Author(s) 2021
M. Hall et al. (eds.), *Chemically Modified Minds*,
https://doi.org/10.1007/978-981-15-6771-1_6

Despite this constant demand for glucose, the brain does not have an in situ store of energy. In comparison to the high energy demand the brain exerts, paradoxically low levels of glucose are stored (as glycogen) within the brain itself. It is estimated that cerebral glycogen stores within the brain would be entirely depleted within ten minutes without continuous circulatory blood glucose supplementation (Benton, 2001), although this may be an optimistic estimation.

Glucose is required not only for energy provision (approximately 30% of the glucose is metabolised through oxidative metabolism to provide energy (Chugani, 1998)), but is also required for the synthesis of amino acids, peptides, lipids and nucleic acids. Glucose is also required to meet the high energy demands of maintaining/restoring ion gradients, for example as required by action potentials, post synaptically, and in the reuptake and recycling of neurotransmitters (Bélanger et al., 2011).

The brain therefore relies almost entirely on circulatory blood glucose crossing the blood-brain barrier. Maintaining homeostasis of blood glucose (through glycaemic control) is of vital importance to adequately fuel the brain, maintaining and protecting brain functioning. When glucose levels are not maintained, or glycaemic control is not effectively functioning, cognitive changes and even neuronal damage can ensue. Fluctuations in acute circulatory glucose even within normative ranges can elicit both enhancements and impairments in a range of cognitive functions.

The mechanisms underpinning the role of glucose on cognition are more complex than we might initially think. In order to understand the mechanisms and the potential impact of glucose as a cognitive enhancer, it is important to first understand how glucose levels are controlled and the processes mediating its availability.

Glycaemic Control

Tightly controlled homeostasis is vital to maintaining and protecting energy provision to the brain. Complex hormonal feedback primarily through the ratio of the hormones glucagon and insulin maintains glycaemic control and restores homeostasis. Both insulin and glucagon are secreted from the pancreas, with opposing actions.

Insulin is secreted from isle of Langerhans β cells in response to elevated blood glucose, typically following feeding (but also in response to other stimuli, e.g. expectation of food). Insulin then stimulates several actions to reduce blood glucose levels. Insulin stimulates the metabolism of glucose to synthesise glycogen via glycogenolysis, which is stored in the liver and muscle. By promoting the use of amino acids in the periphery, insulin acts to downregulate gluconeogenesis (biosynthesis of glucose from non-carbohydrate substrates, e.g. lactate or pyruvate), promoting glycolysis (metabolism of glucose) to meet energetic demands. The presence of insulin also inhibits the secretion of glucagon from the isle of Langerhans α cells, reducing hepatic glucose production and increasing conversion of glucose to fat to be stored in adipose cells. The negative feedback of decreasing blood glucose subsequently inhibits the secretion of insulin to return to the body to a homeostatic state.

Glucagon secretion is stimulated via falling blood glucose levels, stimulating processes to increase blood glucose levels to reinstate homeostasis. The primary action of glucagon is the stimulation of glycogenolysis (conversion of glycogen stores into glucose) in the liver, mobilising glucose into the bloodstream. Glucagon also promotes gluconeogenesis (mobilisation of non-glucose substrates). Through feedback mechanisms, insulin secretion is inhibited along with energy storage actions. Glucagon also upregulates lipid metabolism and ketones to be used as energy in the periphery (e.g. muscles), thereby conserving available glucose for the brain.

Whilst the ratio between insulin and glucagon is the main hormonal regulators of glucose homeostasis, it is by no means this simplistic, for example somatostatin and pancreas polypeptide also have contributory feedback roles in regulating endocrine secretions of the pancreas, although this is beyond the scope of this chapter.

Glucose Metabolism in the Brain

Throughout the body, insulin facilitates the uptake of glucose into cells, but this is not the case in the brain. In the brain, glucose is transported through cell membranes via facilitated diffusion. Facilitative diffusion is useful as it relies on concentration gradients to transport glucose, thus negating the requirement for both insulin and energy. Specific carrier proteins shuttle glucose across the membrane, into the extracellular fluid at a faster rate than diffusion alone would allow. Glucose can then be taken up into the neurons. Glucose is taken up through specific glucose transporters (GLUTs). The concentration of GLUTs varies dependent upon location, with specific transporters adapted to specific tissue needs. GLUT 1, 3 and 4 transporters specifically are found in the brain.

GLUT1 transporters are found in the endothelial cells, allowing glucose to be transported across the blood-brain barrier. The distribution of GLUT1 transporters is skewed, with three–four times as many present on the abluminal (brain side) as opposed to the luminal (blood vessel) side (Farrell & Pardridge, 1991; Messier, 2004). The skewed placement of GLUT1 is thought to be crucial in maintaining a concentration gradient, allowing glucose to diffuse from the blood, into the lower concentrate endothelial intracellular fluid (approximately 20–30% lower than circulatory blood glucose levels) (Messier, 2004). Subsequently glucose is transported into the brain extracellular fluid, from where a considerable proportion is transported (again via GLUT1) into astrocytes. The limited amount of glucose stored as glycogen in the brain is primarily stored in the astrocytes (Swanson, 1992). The glycogen is rapidly turned over, rather than 'stored' per se.

GLUT3 transporters are found on the neurons themselves, transporting glucose into the neuron from the extracellular fluid. GLUT3 allows direct provision of glucose from the blood (via the endothelial cells and extracellular fluid), to be metabolised in the neuron for energy provision. Alternatively, energy is available to the neurons from the astrocytes, which is believed to be primarily transferred as lactate (a product of glycolysis) but can also be released as glucose. Understanding of the cooperation between astrocytes and neurons is still poorly understood and a field of rapidly emerging findings (Bélanger et al., 2011).

The GLUT4 is mediated by insulin and is responsible for insulin-stimulated glucose uptake into skeletal muscle and adipose cells. GLUT4 normally resides in an intracellular membrane compartment, but rapidly populates the plasma membrane in the presence of insulin. This then allows the influx of glucose through facilitative diffusion (McCarthy &

Elmendorf, 2007). The GLUT4 is however also present at specific sites within the brain, for example in the hippocampus, making this area sensitive to insulin. Consequently, insulin does mediate glucose levels within certain areas of the brain (see later discussion for further detail).

Impaired Glycaemic Control and Cognition

Glycaemic control is not stable over the lifespan; rather it declines with normal ageing. Poor glycaemic control is prevalent in older adults (Awad, Gagnon, Desrochers, Tsiakas, & Messier, 2002; Messier, 2004; Parsons & Gold, 1992). Declining cognitive function, memory in particular, is also a feature of ageing, with more complex cognitive tasks appearing to be more susceptible to decline (Balota, Dolan, & Duchek, 2000). Episodic/declarative memory seems to be particularly vulnerable to age-associated decline (Zacks, Hasher, & Li, 2000). In nondiabetic older adults (over 70 years), poor glucoregulation was associated with poorer short-term verbal memory, but increased inhibition and long-term verbal memory.

There are many indices of glucoregulation and little consensus in the literature as to which is the most appropriate to use, with some indices seemingly associated with different element of cognition. In older adults (over 70 years) higher glucose levels following an overnight fast were associated with poor short-term verbal memory. However, 2-hour evoked glucose levels were associated with poorer performance in a divided attention task (Wright et al., 2015).

Elderly individuals with poorer glucoregulatory control show greater memory impairments, relative to matched controls with better glucoregulation (Hall, Gonder-Frederick, Chewning, Silveira, & Gold, 1989; Messier, Tsiakas, Gagnon, & Desrochers, 2010).

Poor glycaemic control is also a feature of several diseases/conditions (e.g. diabetes, obesity, and metabolic syndrome), which present physiological challenges to both health and cognition, especially when not well controlled. Understanding the changes in cognition of such populations informs our understanding of how glucose may underpin cognitive functioning.

Hypoglycaemia and Diabetes

Should glycaemic control be disrupted for extended periods, as for example in acute hypoglycaemia, extensive damage to the brain can occur. Individuals may experience cognitive dysfunction including confusion, discomfort, difficulty concentrating and anxiety in addition to other physiological symptoms such as trembling, sweating, weakness and slurred speech (amongst others). Hypoglycaemia at its most severe can induce coma and even death. Frequent episodes of hypoglycaemia (sometimes referred to as 'cerebral insults') are thought to be associated with impaired cognitive functioning, for example in attention, executive functioning, mental speed, and flexibility. Key populations of interest here are those susceptible to suffering frequent episodes of hypoglycaemia, for example those with type 1 diabetes.

In children with type 1 diabetes, slightly lower overall cognition across a broad range of cognition is reported, including attention, executive functioning, mental flexibility, and psychomotor processing speed (Gaudieri, Chen, Greer, & Holmes, 2008; Northam & Lin, 2010). However, learning and memory are seemingly spared on the whole. Some decrements in learning and memory are reported in children with early onset type 1 diabetes and histories of severe hypoglycaemia (Gaudieri et al., 2008; Kaufman, Epport, Engilman, & Halvorson, 1999; Naguib, Kulinskaya, Lomax, & Garralda, 2009). In adults with type 1 diabetes, these deficits seem to mirror those observed in children. Mild cognitive dysfunctions in the form of decrements to mental speed and flexibility (speed of information processing, psychomotor efficiency, visual and sustained attention, cognitive flexibility and visual perception) are reported with learning and memory spared (see meta-analysis; Brands, Biessels, de Haan, Kappelle, & Kessels, 2005). There are some inconsistencies in the domains reportedly affected, potentially due to differing task constraints used across studies and the degree of damage caused during intensity and frequency of the cerebral insults. Whilst these decrements are typically mild and unlikely to interfere with day-to-day functioning, they may be somewhat magnified in high demand situations. The mechanisms underpinning these decrements are unclear, but may be mediated through microvascular damage as opposed to metabolic control.

In contrast with type 2 diabetes, type 1 is more likely to have much younger onset. The basal metabolic rate in children is also approximately twice that of adults, at least in part driven by the extensive synthesis of new tissue (Kennedy & Sokokloff, 1957). Despite this, early onset of type 1 diabetes does not manifest in more pronounced cognitive decrements than later onset (as in adults with types 2 diabetes) as we might expect in line with greater cerebral insults. Increased neural plasticity in this young population may underpin the potential protection of cognitive functioning, which is not found in adults with later onset type 2 diabetes, for whom neural pathways are already mature.

The literature examining the impact of hypoglycaemia on cognitive functioning in type 2 diabetics is very limited. There does appear to be evidence of an association between severe hypoglycaemia and decrements in cognitive abilities, along with accelerated cognitive decline (Feinkohl et al., 2014). Decrements to processing speed, nonverbal memory, executive function and reasoning were the most commonly observed.

In adults, type 1 diabetes has been shown to slow mental speed and diminish mental flexibility, with impairments ranging from mild to moderate. However, memory and learning remain seemingly intact (Brands et al., 2005). Conversely in type 2 diabetes, memory and learning are targeted in addition to mental speed and flexibility. The majority of diabetic patients suffer impairments in verbal memory that worsen with age and duration of the disease (Elias et al., 1997; Fontbonne, Berr, Ducimetiere, & Alperovitch, 2001; Greenwood, Kaplan, Hebblethwaite, & Jenkins, 2003; Ryan & Geckle, 2000; Strachan, Deary, Ewing, & Frier, 1997). Decrements are also observed in executive functioning, information processing speeds and memory (Awad, Gagnon, & Messier, 2004). Even in the early stages of diabetes, modest cognitive impairments can be observed, specifically in memory (Ruis et al., 2009). Improving the glycaemic control in diabetic patients has been shown to lead to a corresponding improvement in memory tasks (Ryan, 2006).

The decrements in type 2 diabetes seemingly reflect those observed in ageing and other patient populations presenting with impaired glucoregulation and cognition.

Glucose and Impaired Cognition Within Patient Populations

Poor glucoregulatory control is a key feature of several conditions presenting with severe cognitive decrements. Here we briefly consider the pathology of the conditions, which informs understanding of the potential mechanisms underlying the glucose facilitation effect. The specific glucose-induced cognitive enhancements that have been demonstrated in these populations are also discussed.

Dementia

Alzheimer's disease (AD) is a neurodegenerative disease, presenting with progressively deteriorating brain functions including memory, understanding, judgement, language, and thinking (Luengo-Fernandez, Leal, & Gray, 2010). Impaired glucoregulation is a feature of the disease, with diabetes a high-risk factor across several types of dementia, including Alzheimer's and vascular types (Biessels, Staekenborg, Brunner, Brayne, & Scheltens, 2006). Several other risk factors are also associated with poor glucoregulation: abnormal lipid metabolism, oxidative stress, inactive lifestyles, obesity, and decreased cerebral blood flow (Martins et al., 2006).

Glucose has been shown to enhance memory in this population, or more accurately, performance decrements are attenuated in this impaired group, rather than returning cognitive functioning to those of a healthy individual. In patients with moderate to severe probable senile AD, a (75 g) glucose load has been shown to facilitate memory and a range of cognitions including orientation, narrative prose, face recognition, word recognition, and recall (Manning, Ragozzino, & Gold, 1993).

Patients with mild senile AD and poorer glucoregulatory control (but not better glucoregulatory control) benefited from glucose facilitation when completing a verbal episodic memory (paragraph recall task). Conversely, in matched controls, facilitation was observed only in those with better (not poorer) glucoregulation (Craft, Zallen, & Baker, 1992). In a further study Craft et al. reported glucose memory facilitation in subjects with very mild AD. In this study glucose was elevated using a hyperglycaemic clamp to maintain plasma glucose levels of 9.7 mmol/L or 12.5 mmol/L. Relative to fasting performance, verbal episodic memory was enhanced (at the higher glucose elevation), however this effect was not elicited at follow-up 18 months later in subjects whose AD had progressed (Craft et al., 1993).

These AD studies (Craft et al., 1992, 1993; Manning et al., 1993) provide supporting evidence that memory is mediated by glucoregulatory processes, particularly in individuals with reduced brain glucose metabolism. Interesting glucose appears to be an effective short-term intervention for improving cognitive functioning during the very early stages of decline, at least for some patients.

Cerebrovascular disease and declining cerebral metabolism deficits are evident in AD, which reflect the neuro-degeneration within this disease (Messier, 2003). It is thought the behavioural impairments stem from cholinergic degeneration in the basal forebrain and hippocampal formation, accounting for memory decline (Craft & Watson, 2004). Modulation of cholinergic processes in the brain is one potential mechanism responsible for the glucose memory enhancement. This facilitation of memory in the early stages of AD supports the postulation that glucose may mediate memory processes via modulation of cholinergic activity.

Schizophrenia

In addition to displaying abnormalities in the perception of reality, schizophrenic patients also display cognitive impairments in learning and memory (Gruzelier et al., 1988), in attention, and in executive functions (Goldberg et al., 1987; Seidman et al., 1991). The most persistent of these deficits is in long-term declarative memory (Stone, Seidman, Wojcik, & Green, 2003). Schizophrenic patients present with a greater risk of obesity, diabetes, smoking, poor dietary habits and the use of antipsychotic medication, all of which are detrimental to glucoregulation (Maric et al., 2008). Consequently impaired glucose tolerance and insulin resistance are common features of schizophrenia (Fucetola, Newcomer, Craft, & Melson, 1999; Schultz, Arndt, Ho, Oliver, & Andreasen, 1999).

Glucose has been found to enhance verbal declarative memory (Fucetola et al., 1999; Newcomer et al., 1999; Stone et al., 2003) but to also impair vigilance/attention (Fucetola et al., 1999; Stone et al., 2003) in this population. A functional magnetic resonance imaging (fMRI) study (Stone, Thermenos, Tarbox, Poldrack, & Seidman, 2005) revealed increased activation of the parahippocampal regions in schizophrenic patients during encoding following a glucose load, although memory was not found to be enhanced in this study.

Stone and Hsi (2011) postulate that it is likely that similar underlying aetiology is driving the deficits observed in declarative memory and hippocampal abnormalities in schizophrenia, as is responsible for deficits in other pathologies presenting with poor glucoregulation/availability of glucose.

Glucose Enhancement of Cognition

Cognition is inextricably linked with circulatory blood glucose levels, although the underlying physiology is complex. Glycaemic control is tightly controlled with minimal variation in circulatory blood glucose, and optimal levels quickly restored following a glucose challenge (in healthy individuals). It seems somewhat counterintuitive then, that fluctuations in blood glucose within the normal range can exert changes in cognitive functioning. However, this is the case, even within healthy populations whereby administration of a glucose load can and does mediate cognitive functioning.

Acute changes in blood glucose levels through glucose consumption have been shown to directly modulate cognitive functioning, with specific modalities seemingly more or less susceptible. Often referred to as the 'glucose facilitation effect', the glucose enhancement of cognition has become widely reported and accepted over the last 30 years or so (Messier, 2004). The glucose facilitation effect has been reported in both healthy human and animal studies of younger and older individuals, in addition to populations with cognitive impairments (e.g. Schizophrenia and Dementia). Learning and memory are seemingly predominantly susceptible to facilitation, with declarative memory widely reported to be improved following a glucose load. This finding is well accepted and reported across the literature (for reviews, see Benton, 2001; Lieberman, 2003; Messier, 2004; Riby, 2004; Smith et al., 2011a). There are however inconsistencies within the literature, with several studies reporting no facilitation following a glucose load in healthy individuals, potentially due to mitigating factors (for discussion, see Hoyland, Lawton, & Dye, 2008).

Acute Administration and Cognition

Several studies have investigated experimentally the impact of acute (short-term) elevation of glucose levels on cognition. Elevating circulatory blood glucose levels is straight forward. Typically, a glucose treatment in drink form is administered to participants. The specific treatment composition varies across studies but is most frequently comprised of powdered glucose (25 g and 50 g are the most common doses), dissolved in water (typically 150–300 ml) with a flavour mask (orange cordial or lemon juice). Placebo drinks are typically matched for sweetness, with either saccharine or aspartame non-nutritive sweeteners. A dose of 25 g glucose has been shown to reliably elevate circulating blood glucose levels following a 10–15 minute absorption period, at which point cognitive testing is then completed. Studies utilise a mix of between and within participant designs, with some studies specifically measuring glucoregulatory control (e.g. through an oral glucose tolerance test). Participants are normally tested following a fasted period, either an overnight 12-hour fast or a 2-hour fast.

As with many nutritional and pharmaceutical treatments, glucose has been shown to have an inverted 'U' dose-response curve (Parsons & Gold, 1992). A dose of 25 g of glucose has been shown to optimal dose for inducing memory enhancement in healthy young adults (Leigh M. Riby, 2004) and healthy adolescents (Smith & Foster, 2008). However, both 25 g and 50 g doses of glucose have been reported to be optimal in healthy elderly humans (Hall et al., 1989; Manning, Stone, Korol, & Gold, 1998; Parsons & Gold, 1992; Riby et al., 2009). Optimal dosage may, however, be mediated through the period of fasting prior to testing.

Owen et al. (2012) reported 25 g glucose elicited performance enhancements following a 2-hour pre-test fast, but that 60 g glucose was effective following a 12-hour fast. A 60 g glucose dose was also more effective in enhancing implicit memory (Owen, Finnegan, Hu, Scholey, & Sünram-Lea, 2010). In human studies, these standardised dosages are employed, however body composition difference between participants will cause variance in the actual blood glucose concentration, as will individual glucoregulatory control.

Declarative Memory

Declarative memory is the most consistently reported cognition to benefit from elevated glucose (Hoyland et al., 2008; Messier, 2004; Smith et al., 2011a). Primarily verbal declarative/episodic memory tasks, for example paragraph recall, word recall and word recognition, have been widely reported to be sensitive to glucose facilitation (Benton, Owens, & Parker, 1994; Meikle, Riby, & Stollery, 2004; Messier, Pierre, Desrochers, & Gravel, 1998; Owen et al., 2010; Riby et al., 2006; Scholey et al., 2013; Sünram-Lea, Foster, Durlach, & Perez, 2001).

Declarative memory relies on the hippocampus. As such several proposed mechanisms of action for the glucose facilitation effect suggest that it is mediated through the hippocampal formation (see discussion of proposed mechanisms underlying the 'hippocampus hypothesis'). Studies attempting to explore this postulation have explored how glucose may mediate familiarity and recollection components of memory using the Remember-Know paradigm. Recollection (remember) is thought to be mediated by the hippocampus, with memories associated with episodic richness and explicit recall of spatio-temporal contextual details of the event. Familiarity (know), however, lacks this episodic richness and contextual details, leaving a 'feeling of knowing'. Familiarity is thought to be mediated via non-hippocampal structures such as the perirhinal cortex (Aggleton & Brown, 2006). The findings thus far have been somewhat contradictory. Glucose has been shown to specifically enhance hippocampally mediated recollection memory (Riby et al., 2008; Smith, Riby,

Sünram-Lea, van Eekelen, & Foster, 2009); however, this is refuted in other studies reporting wider (recollection and familiarity) facilitation, rather than specific hippocampal mediation (Scholey et al., 2013; Smith et al., 2009).

An alternative task, the Process Dissociation Procedure (PDP) task, is a more cognitively demanding paradigm that also distinguishes between hippocampally driven recollection and non-hippocampally driven familiarity (Jacoby, 1991). Participants make a recognition judgement, followed by an include/exclude decision of 'old' items based on a categorisation (e.g. rejecting items at recognition that were rated as 'pleasant' during encoding). This requires participants to access contextual details of memory and make a decision on that basis. Glucose facilitated performance, reducing the number of failures to exclude items (highly demand) in comparison to placebo (Brandt, 2015). This suggests glucose is specifically mediating hippocampally driven memory during high cognitive demand.

Parent et al. (2011) further support that glucose is seemingly targeting memory, with an fMRI study showing increased activation in regions and functional connectivity associated with effective encoding (e.g. hippocampus, prefrontal cortex and others) and recall (e.g. hippocampus and amygdala) in response to emotional stimuli.

There is increasing evidence that poorer glucoregulatory control (in subclinical 'healthy' populations) is a mediating factor as to whether a glucose facilitation effect will be observed. Poorer regulation was associated with declarative performance enhancements (improved word recall accuracy) following a 25 g glucose load, with the same dose impairing accuracy in better regulators (Owen, Scholey, Finnegan, & Sünram-Lea, 2013). These conflicting effects may also begin to explain why the glucose facilitation effect is not robustly reported, particularly in studies not controlling for or examining glucoregulatory indices.

Demand Dependent

Decrements in cognition are most easily detected in complex and challenging cognitive tasks, particularly when decrements are subtle, for example in ageing (McNay, 2005). With basic cognitive functions seemingly spared, subtle cognitive decline is often masked until such a time as deficits reach an advanced stage. What constitutes a challenging cognitive task very much depends on the cognitive ability of the individual/population of interest. There is a wide range of factors that mediate performance, some offering a protective effect, for example education and physical activity level. The 'challenge' refers to a task at which the cognitive load is increased to a point whereby performance is compromised. Such increases in cognitive load exert increased metabolic demand resulting in a shortage of oxygen/glucose and therefore metabolic stress. Through increasing the cognitive challenge or demand characteristics of a given cognitive task, greater performance variances begin to emerge, even in healthy young adults.

Rather than a blanket improvement in cognitive functioning, it seems that task demand characteristics are an important moderator of susceptibility to glucose enhancement. The importance of task difficulty versus task domain in the glucose literature is a question that has underpinned the interpretation of several studies and is the basis of two proposed mechanisms that may underlie the facilitation effect (discussed later).

Demand characteristics can be manipulated in a variety of ways, several of which have successfully elicited a glucose facilitation though increasing mental effort/demand. One approach is to employ a battery of tasks with increasing difficulty, for example, glucose facilitated performance in a demanding serial sevens subtraction task (participants are asked to count down from a three-digit number in sevens) but not a lesser demanding serial threes subtraction version of the task (Kennedy & Scholey, 2000). Other studies have utilised prolonged periods task repetition, requiring sustained cognitive demand and glucose utilisation (Kennedy & Scholey, 2000; Owens, Parker, & Benton, 1997; Scholey, Harper, & Kennedy, 2001; Scholey, Laing, & Kennedy, 2006). Alternative approaches have manipulated the difficulty of the stimulus during the tasks, for example through increasing the number of stimuli in memory tasks or manipulating the stimuli (e.g. concrete vs. abstract words), or high versus low imagery stimuli (Meikle, Riby, & Stollery, 2005; Riby et al., 2006; Riby, Meikle, & Glover, 2004).

The employment of secondary 'dual-tasking paradigms' has yielded some interesting findings—this has proven to be an effective strategy to elicit glucose enhancement effects. Primarily completed during encoding phases of memory tasks, a range of secondary tasks have utilised. Asking participants to complete alternating hand movement sequences (requiring the monitoring of which sequence was last completed and then switching to the alternative), whilst encoding has been shown to induce memory performance benefits of glucose (Foster, Lidder, & Sünram, 1998; Scholey et al., 2013; Scholey, Sünram-Lea, Greer, Elliott, & Kennedy, 2009a, 2009b; Sünram-Lea, Foster, Durlach, & Perez, 2002a, 2002b). A secondary card sorting task during episodic remembering did not elicit a glucose enhancement effect (Riby et al., 2004). Performance on the secondary task is not always assessed, often it is simply monitored to check for adherence. Glucose facilitation/mitigation of the secondary task is missed in this case. Additionally, should the secondary task be too difficult, it may ameliorate any performance effects, detracting too much attention from the principal task. Scholey et al. (2009a) reported glucose improvements of a secondary tracking task during memory encoding following glucose, in a dual-tasking paradigm in healthy young adults. Memory in this study was not enhanced, although glucose seemed to act as a cognitive enhancer of attention. Subsequently, glucose effects may be influential on the secondary task, but depending on the task and study design, effects may be being missed. This finding was replicated in healthy older adults (but not younger), with speeded reaction times for recognition (Macpherson et al., 2015).

Although increasing cognitive demand does not automatically exaggerate any existing glucose facilitation effects (Riby et al., 2006), knowledge of glucose as a cognitive enhancer during high levels of cognitive demand has real-world implications and potential applications.

Other Cognitive Domains

The influence of glucose enhancement on a wide range of cognitive domains has been explored in conjunction with declarative memory; however, the evidence for a clear and consistent effect is less convincing than for other domains. For example, glucose has been found to interact with attentional processes (Riby et al., 2008) and to improve attention following a glucose load (Benton, 1990; Meikle et al., 2004). Improvements in working memory visuospatial memory (Sünram-Lea et al., 2001; Sünram-Lea et al., 2002a, 2002b), verbal fluency (Donohoe & Benton, 1999) and face recognition (Metzger, 2000; Metzger & Flint, 2003) have all been reported.

Glucose also seemingly facilitates frontal lobe-mediated inhibition tasks, for example in a Stroop Task (Brandt, Gibson, & Rackie, 2013; Gagnon, Greenwood, & Bherer, 2010), generally through improved reaction times rather than accuracy improvements. However, other sensorimotor tasks appear to be slowed following a glucose load. Glucose slowed response times, but not accuracy in the Eriksen flanker task which evokes cognitive conflict (Hope, Seiss, Dean, Williams, & Sterr, 2013).

There has been some interesting recent work on inhibition processes and self-control in relation to glucose and glucoregulatory control. Tentative evidence has suggested that behavioural flexibility, the ability to adapt and modify behaviour in response to environmental demand may be susceptible. Glucose supplementation was found to reduce impulse related choice behaviour in seemingly better glucoregulators (Riby et al., 2017). However, the fasting and dosing strategy within this study is not comparable with the majority of the literature, with participants tested post-prandially and given a much lower dose of glucose (15 g as opposed to the more standard 25 g or 50 g). The study does however provide some interesting forays into the potential influence of glucose on cognition within a more naturalistic physiological state.

Self-Control and Inhibition

Self-control and inhibition processes are important in the regulation of behaviour and decision-making in everyday life. These processes require effortful, controlled, executive functions, which are more metabolically costly than automatic processes (Gailliot et al., 2007). Recent findings suggest that these processes may be moderated by glucose, which has important ramifications beyond lab-based cognitive testing. Self-control findings fit well with the findings of glucose on executive functioning. In an extended Stroop task (45 minutes), incongruent colour Stroop elicited reduced blood glucose levels, suggesting the mental effort required in the executive control of inhibition of an automatic response, increases energy mobilisation and utilisation (in the absence of a glucose load), although falling blood glucose levels did not mediate performance decrements (Fairclough & Houston, 2004).

Self-control is seemingly limited by the provision of energy in the form of glucose. Should these supplies be depleted, for example through a demanding self-control task, subsequent self-control and inhibition is reduced (Gailliot et al., 2007). The classic and very much real-world example being resisting eating an enticing snack (freshly baked cookie) subsequently reduces participants' persistence threshold in an effortful follow-up task. With an increased tendency for participants to give up faster than if they had not engaged in the self-control/willpower task. This effect has been replicated in a variety of contexts; for example, conscious emotion regulation and thought suppression have been shown to reduce subsequent performance on tasks requiring self-control and willpower (Muraven, Tice, & Baumeister, 1998).

An investigation by Gailliot et al. (2007) postulated that, while metaphorically self-control has been described as being a limited resource, there may be grounding for this metaphor in terms of limited physiological resources. Galliot et al. reported acts of self-control reduced blood glucose levels, which in turn predicted poor performance on follow-up self-control tasks. Furthermore, administration of a glucose drink ameliorated the decline in self-control in the follow-up task. This finding raises intriguing forays into the potential cognitive impact of glucose and its role in everyday behaviours, emotion, societal interactions, and functioning. For example, self-control in food choices in dieters and subsequent disinhibition in food choices.

Acute administration of glucose has been found to facilitate executive functioning, reducing aggression in response to provocation in participants scoring high for trait aggression (following a self-control depletion task). However, glucose was also found to increase aggression when not provoked, again in participants scoring high for trait aggression (Denson, von Hippel, Kemp, & Teo, 2010).

Such studies, whilst few in number, can begin to explain how glucose and glucoregulatory control may mediate everyday decision-making and performance.

Glucoregulation

Increasing evidence suggests that glucose enhancement may be preferentially enhancing cognition in those with poorer glucoregulatory control, even within normal subclinical parameters. As previously discussed, in older adults, poor glucoregulation is associated with declining cognitive functioning, with glucose enhancing cognition in the poorer but not better regulators.

Rather concerning are the decrements in performance that are reported even in healthy young adults with poorer (but not clinically impaired) glucoregulatory control (Awad et al., 2004; Lamport, Lawton, Mansfield, & Dye, 2009). Young, better glucoregulators have been shown to outperform young poor regulators, on a range of cognitive tasks, for example memory (e.g. Awad et al., 2002; Donohoe & Benton, 2000; Messier, Desrochers, & Gagnon, 1999; Owen et al., 2013) and attention (Donohoe & Benton, 1999). Furthermore, evoked blood glucose as a measure of glucoregulatory control has been shown to be associated with verbal memory performance in young adults (Messier, Awad-Shimoon, Gagnon, Desrochers, & Tsiakas, 2011). These worrying findings suggest that mild cognitive decline associated with poor glycaemic control is in evidence long before this becomes apparent in everyday life and prior to the onset of impaired glucose regulation.

Participants were assessed after drinking glucose or saccharin, using a repeated-measures design. There was no effect of glucose on cognitive performance. Glucoregulatory indices calculated based on insulin levels or fasting glucose levels, explained less cognitive variability compared to indices based on evoked glucose levels. These findings suggest that cognitive decrements are observable in young, nondiabetic adults, prior to the onset of impaired glucose regulation and diabetes.

Even the relatively small differences in better and poorer glucoregulatory control in healthy young adults seemingly moderate the influence glucose administration has on subsequent performance. It has been suggested that glucose preferentially facilitates memory, where memory is operating suboptimally (Messier, 2004) and that suboptimal glucoregulation underpins that decline in memory. Glucose has been found to enhance performance in poorer but not better regulators, following a glucose load in declarative memory tasks (free word recall, paragraph recall, word order recall) (Messier et al., 1999), a finding that replicates those observed in older adults. Tentative evidence suggests that glucose may impair performance in better (healthy young) glucoregulators. Owen et al. (2013) reported a 25 g glucose load enhancing word recall accuracy in poorer regulators but impairing accuracy in better regulators, a finding that has also been observed in older samples. This is an area of growth within the literature, with few studies to date including objective standalone measures of glucose regulation through glucose tolerance tests.

Other Factors

There is an almost inexhaustible list of factors that may influence glucose enhancement of cognition (memory, attention, or other) in any given individual. It is pertinent to consider the medium through which glucose is typically administered. The glucose and placebo drinks, although matched for sweetness and mouthfeel, have very different hydrating properties. Scholey et al. (2009b) reported declarative memory enhancements following a glucose drink in participants reporting low initial thirst, but memory decrements in those presenting with high thirst. Other factors include age, gender and even anxiety (Riby et al., 2009; Smith et al., 2011a).

Mechanisms of Effect

The specific mechanisms underpinning the glucose enhancement effect are yet to be isolated and potentially involve a complicated interaction of effects. Several theories have been proposed which are not mutually exclusive. Each theory is seemingly more or less likely to be influential in underpinning glucose/glucoregulatory effects, dependent upon the specific element of cognition being considered. Broadly, the theories fall into two approaches: global or domain specific. Various mechanisms propose that raising glucose levels lead to glucose acting directly on the brain by altering neural metabolism, neural activity, and/or neurotransmitter synthesis (Korol & Gold, 1998). Alternative approaches suggest that it may be peripheral processes/organs that mediate the glucose effect on cognition, for example the liver or insulin effects (White, 1991). Many of the key elements of the proposed mechanisms have been touched upon already, however they are more specifically addressed here along with brief examples of supporting evidence.

Global Energy Provision

The most intuitive proposed method of glucose enhancement on cognition is the elevation of circulatory glucose levels providing increased availability of energy to the brain, which in turn facilitates cognitive enhancements. This is a neat theory with a clear path between energy provision and performance, seemingly a common-sense theory.

Cognitively demanding tasks in humans have been shown to be associated with depleted circulating glucose (Donohoe & Benton, 1999; Fairclough & Houston, 2004; Scholey et al., 2001, 2006), with glucose administration replenishing the available glucose, thereby preventing energy shortfalls and facilitating cognitive enhancements/ameliorating decrements. However, this simplistic approach does not explain the inverted 'U' dose-response curve.

Glucose levels in the extracellular fluid across the brain are not uniform, nor as stable as traditionally believed. Dissociations have been observed between fluctuating blood glucose levels and extracellular glucose levels within the brain, highlighting that simply increasing circulatory glucose does not automatically increase availability. McNay's work in rats (e.g. McNay, McCarty, & Gold, 2001) suggests that increased glucose supply is specifically mediating brain areas involved in performance, for example the hippocampus, with these areas subsequently benefiting from increased glucose availability (restoring levels, rather than elevating) as opposed to a generalised beneficial impact across all brain structures.

The Hippocampus Hypothesis

The postulation that the hippocampus underlies the glucose facilitation of memory has been referred to as the domain approach. The hippocampus is strongly implicated in memory processes and is sensitive to both glucose availability and depletion. Again, this hypothesis appears sensible, however the route(s) through which this increased hippocampal functioning may be facilitated remains unclear and subject to conjecture.

In rodent studies, direct infusions into specific brain regions, such as the hippocampus, have shown a restorative effect on drug induced memory deficits (Canal, 2005; Krebs-Kraft & Parent, 2008; McNay et al., 2001; McNay & Gold, 2001; Ragozzino & Gold, 1995; Ragozzino, Pal, Unick, Stefani, & Gold, 1998; Stefani, Nicholson, & Gold, 1999). For example, administration of glucose reversed the age-associated memory performance decrements and ameliorated the glucose depletion observed in the hippocampus in rats (McNay & Gold, 2001). These studies provide convincing evidence that the glucose facilitation effect may be specifically targeting the hippocampus, whilst also elucidating why glucose enhancement may be predominantly observed in poorer glucoregulators. This and similar findings (See Gold, 2014 for an insightful review) provide compelling evidence that the hippocampus is specifically implicated in the glucose enhancement effect.

In humans, memory recognition paradigms have been employed to attempt to dissociate whether glucose is specifically mediating hippocampal function and episodic memory (Brandt, 2015; Scholey et al., 2013; Smith et al., 2009; Sünram-Lea, Dewhurst, & Foster, 2008). However, these behavioural studies have reported mixed findings. Imaging studies have demonstrated increased activation of the parahippocampal cortex during encoding, but no memory enhancements in schizophrenic patients following glucose (Stone et al., 2005). In healthy adults, increased activation associated with episodic memory in the hippocampus and prefrontal cortex with memory improvements, lends support to the hippocampus hypothesis (Parent et al., 2011).

The evidence in both human and animal studies discussed suggests that the hippocampus specifically is, at least partially, underlying the glucose facilitation effect.

Insulin

The hippocampus is one of the few areas of the brain with an abundance of insulin receptors and insulin-sensitive GLUT4 glucose transporters in situ (Messier, 2004). Hippocampal glucose metabolism is therefore sensitive to peripheral insulin levels in addition to circulatory glucose levels. Not only does insulin promote glucose uptake into the hippocampus, but insulin has also been shown to act as a cognitive enhancer (Watson & Craft, 2004).

Craft et al. (1996) reported that the memory effects of hyperglycaemia in AD were replicated in clamping studies which induced hyperinsulinemia whilst maintaining fasting glucose baseline levels (note that glucose was also administered to prevent hypoglycaemia). The firing rate in the hippocampus has been shown to be sensitive to insulin (Hoyer, 2003). Insulin administration has also been shown to facilitate memory in both elderly (Reger et al., 2006) and early AD patients (Reger et al., 2008). These populations both have a high risk of poor/declining glucoregulation and likely, some degree of insulin resistance. This lends further support to insulin as a neuroendocrine, may be underpinning the glucose facilitation effect. Insulin and glucose levels are intrinsically linked, with no foolproof way of investigating the effect of one without moderating the other.

Hippocampal Acetylcholine (ACh) Synthesis

Several key neurotransmitters within the brain, including ACh, glutamate, and GABA, rely directly on the provision of circulatory glucose for their synthesis (Messier, 2004). The elevation of circulatory glucose levels may facilitate hippocampal function (and therefore memory), through increasing cholinergic activity (Messier, 2004; Messier, Gagnon, & Knott, 1997). The metabolism of glucose provides the substrate for synthesis of acetyl-CoA, a precursor for the formation of the acetylcholine, a cholinergic receptor agonist. Psychopharmacological manipulation of cholinergic agonists and antagonists has been used to replicate fMRI brain activity patterns observed in ageing and lend support to the cholinergic compensating approach to treating cognitive deficits in mild AD (Dumas & Newhouse, 2011). Cholinergic agonists have also been used to attenuate the amnesic effects of scopolamine, through increased ACh (Durkin, Messier, de Boer, & Westerink, 1992; Messier, 1998). The reproduction of brain activity patterns through cholinergic agonists/antagonists, combined with the facilitatory effects of cholinergic agonists on memory, provides some convincing evidence for the hippocampal acetylcholine synthesis mechanism of effect.

It is likely that several of the proposed mechanisms (and several not discussed here) are having a causative effect.

Can We Use Exploit Glucose/Glycaemic Control to Enhance Cognition?

This a tricky question to answer responsibly, but ultimately yes, we can. Glucose is already being exploited to improve ergogenic performance in athletes, as a self-treatment for fatigue, with attempts to market the cognitive performance-enhancing properties of glucose within products brought to market (supplements such as ginseng/guarana with glucose, caffeine and glucose, chocolate boost bars, and energy drinks). Glucose holds little monetary value from a pharmaceutical perspective but it is readily available to purchase (dextrose powder/glucose tablets).

It is feasible to exploit the glucose enhancement effects to enhance cognition short term, with the caveat that repeated hyperglycaemia would be detrimental longer term to glucoregulatory control, as evidenced by the implication of poor glycaemic control in cognitive decline. However, understanding the role of glucose and glycaemic control on cognition offers the intriguing possibility to exploit these mechanisms responsibly.

Through understanding the impact of self-control in reducing blood glucose levels and subsequent self-control/willpower, we gain an insightful understanding of everyday behaviours. Knowing that a glucose load (or consumption of a calorific source to raise blood glucose) reinstates self-control hints at a powerful tool to moderate everyday behaviour, for example, in assisting individuals to make better food choices, ultimately enhancing health and mitigating risk factors associated with declining glucoregulatory disorders. Strict dieting requires continuous self-control and willpower, often to no avail, with overeating on calorie-dense food later in the day, effectively sabotaging any calorific deficit. This knowledge can allow us to develop more effective strategies to manage cognitive self-control and decision-making.

Glucose is particularly effective in enhancing (or moderating deficits in) declarative memory, specifically during high cognitive demand. Often, we can anticipate situations in which we will encounter significant cognitive demand and declarative memory, for example job interviews, exams, public speaking, busy work periods. Effectively managing glucose levels within such situations can facilitate our performance.

The effect of glucose is far from straight forwards, with many moderating factors. What we do know, is that suboptimal circulatory glucose levels are detrimental to cognitive functioning, even in healthy populations. So, while there is scope to utilise elevated glucose as a cognitive enhancer (short term), a more responsible approach is to effectively manage glucoregulatory control long term to mitigate risk factors associated with conditions presenting with cognitive impairments and cognitive decline.

Ensuring maximally efficient glucoregulation should be a high priority both for individuals and within public policy. Focusing on maintaining effective glycaemic control is crucial in mediating risk factors that are associated with cognitive decline and brain ageing (e.g. obesity, diabetes, dementia). Eating healthily, minimising obesity, and increasing exercise are ultimately the most desirable ways to maintain efficient glucoregulation and mediate the associated risk factors for cognitive decline.

References

Aggleton, J. P., & Brown, M. W. (2006). Interleaving brain systems for episodic and recognition memory. *Trends in Cognitive Sciences, 10*(10), 455–463.

Awad, N., Gagnon, M., Desrochers, A., Tsiakas, M., & Messier, C. (2002). Impact of peripheral glucoregulation on memory. *Behavioral Neuroscience, 116*(4), 691–702.

Awad, N., Gagnon, M., & Messier, C. (2004). *The relationship between impaired glucose tolerance, type 2 diabetes, and cognitive function. Journal of Clinical and Experimental Neuropsychology, 26*, 1044–1080.

Balota, D. A., Dolan, P. O., & Duchek, J. M. (2000). Memory changes in healthy older adults. In *The Oxford handbook of memory* (pp. 395–409). Oxford University Press.

Bélanger, M., Allaman, I., & Magistretti, P. J. (2011). Brain energy metabolism: Focus on Astrocyte-neuron metabolic cooperation. *Cell Metabolism, 14*(6), 724–738.

Benton, D. (1990). The impact of increasing blood glucose on psychological functioning. *Biological Psychology, 30*(1), 13–19.

Benton, D. (2001). The impact of the supply of glucose to the brain on mood and memory. *Nutrition Reviews, 59*, S20.

Benton, D., Owens, D. S., & Parker, P. Y. (1994). Blood glucose influences memory and attention in young adults. *Neuropsychologia, 32*(5), 595–607.

Biessels, G. J., Staekenborg, S., Brunner, E., Brayne, C., & Scheltens, P. (2006). Risk of dementia in diabetes mellitus: A systematic review. *Lancet Neurology, 5*(1), 64–74.

Brands, A. M. A., Biessels, G. J., de Haan, E. H. F., Kappelle, L. J., & Kessels, R. P. C. (2005). The effects of type 1 diabetes on cognitive performance: A meta-analysis. *Diabetes Care, 28*(3), 726–735.

Brandt, K. R. (2015). Effects of glucose administration on category exclusion recognition. *Journal of Psychopharmacology, 29*(7), 777–782.

Brandt, K. R., Gibson, E. L., & Rackie, J. M. (2013). Differential facilitative effects of glucose administration on stroop task conditions. *Behavioral Neuroscience, 127*(6), 932–935.

Canal, C. E. (2005). Glucose injections into the dorsal hippocampus or dorsolateral striatum of rats prior to T-maze training: Modulation of learning rates and strategy selection. *Learning & Memory, 12*(4), 367–374.

Chugani, H. T. (1998). A critical period of brain development: Studies of cerebral glucose utilization with PET. *Preventive Medicine, 27*(2), 184–188.

Craft, S., Dagogo-jack, S. E., Wiethop, B. V., Murphy, C., Nevins, R. T., Fleischman, S., … Cryer, P. E. (1993). Effects of hyperglycemia on memory and hormone levels in dementia of the Alzheimer type: A longitudinal study potentially interfering with these processes and making older. *Behavioral Neuroscience, 107*(6), 926–940.

Craft, S., Newcomer, J., Kanne, S., Dagogo-Jack, S., Cryer, P., Sheline, Y., … Alderson, A. (1996). Memory improvement following induced hyperinsulinemia in Alzheimer's disease. *Neurobiology of Aging, 17*(1), 123–130.

Craft, S., & Watson, G. S. (2004). Insulin and neurodegenerative disease: Shared and specific mechanisms. *Lancet Neurology, 3*(3), 169–178.

Craft, S., Zallen, G., & Baker, L. D. (1992). Glucose and memory in mild senile dementia of the Alzheimer type. *Journal of Clinical and Experimental Neuropsychology, 14*(2), 253–267.

Denson, T. F., von Hippel, W., Kemp, R. I., & Teo, L. S. (2010). Glucose consumption decreases impulsive aggression in response to provocation in aggressive individuals. *Journal of Experimental Social Psychology, 46*(6), 1023–1028.

Donohoe, R. T., & Benton, D. (1999). Cognitive functioning is susceptible to the level of blood glucose. *Psychopharmacology, 145*(4), 378–385.

Donohoe, R. T., & Benton, D. (2000). Glucose tolerance predicts performance on tests of memory and cognition. *Physiology and Behavior, 71*(3–4), 395–401.

Dumas, J. A., & Newhouse, P. A. (2011). The cholinergic hypothesis of cognitive aging revisited again: Cholinergic functional compensation. *Pharmacology, Biochemistry, and Behavior, 99*, 254–261.

Durkin, T. P., Messier, C., de Boer, P., & Westerink, B. H. C. (1992). Raised glucose levels enhance scopolamine-induced acetylcholine overflow from the hippocampus: An in vivo microdialysis study in the rat. *Behavioural Brain Research, 49*(2), 181–188.

Elias, P. K., Elias, M. F., D'Agostino, R. B., Cupples, L. A., Wilson, P. W., Silbershatz, H., & Wolf, P. A. (1997). NIDDM and blood pressure as risk factors for poor cognitive performance. The Framingham study. *Diabetes Care, 20*(9), 1388–1395.

Fairclough, S. H., & Houston, K. (2004). A metabolic measure of mental effort. *Biological Psychology, 66*(2), 177–190.

Farrell, C. L., & Pardridge, W. M. (1991). Blood-brain barrier glucose transporter is asymmetrically distributed on brain capillary endothelial lumenal and ablumenal membranes: An electron microscopic immunogold study. *Proceedings of the National Academy of Sciences of the United States of America, 88*(13), 5779–5783.

Feinkohl, I., Aung, P. P., Keller, M., Robertson, C. M., Morling, J. R., McLachlan, S., … Price, J. F. (2014). Severe hypoglycemia and cognitive decline in older people with type 2 diabetes: The edinburgh type 2 diabetes study. *Diabetes Care, 37*(2), 507–515.

Fontbonne, A., Berr, C., Ducimetiere, P., & Alperovitch, A. (2001). Changes in cognitive abilities over a 4-year period are unfavorably affected in elderly diabetic subjects: Results of the epidemiology of vascular aging study. *Diabetes Care, 24*(2), 366–370.

Foster, J. K., Lidder, P. G., & Sünram, S. I. (1998). Glucose and memory: Fractionation of enhancement effects? *Psychopharmacology, 137*(3), 259–270.

Fucetola, R., Newcomer, J. W., Craft, S., & Melson, A. K. (1999). Age- and dose-dependent glucose-induced increases in memory and attention in schizophrenia. *Psychiatry Research, 88*(1), 1–13.

Gagnon, C., Greenwood, C. E., & Bherer, L. (2010). The acute effects of glucose ingestion on attentional control in fasting healthy older adults. *Psychopharmacology, 211*(3), 337–346.

Gailliot, M. T., Baumeister, R. F., Dewall, C. N., Maner, J. K., Plant, E. A., Tice, D. M., … Schmeichel, B. J. (2007). Self-control relies on glucose as a limited energy source: Willpower is more than a metaphor. *Journal of Personality and Social Psychology, 92*(2), 325–336.

Gaudieri, P. A., Chen, R., Greer, T. F., & Holmes, C. S. (2008). Cognitive function in children with type 1 diabetes. *Diabetes Care, 31*(9), 1892–1897.

Goldberg, T. E., Weinberger, D. R., Berman, K. F., Pliskin, N. H. & Podd, M. H. (1987). Further Evidence for Dementia of the Prefrontal Type in Schizophrenia–a Controlled-Study of Teaching the Wisconsin Card Sorting Test. *Archives of General Psychiatry, 44*, 1008–1014.

Gold, P. E. (2014). Regulation of memory—From the adrenal medulla to liver to astrocytes to neurons. *Brain Research Bulletin, 105*, 25–35.

Greenwood, C. E., Kaplan, R. J., Hebblethwaite, S., & Jenkins, D. J. (2003). Carbohydrate-induced memory impairment in adults with type 2 diabetes. *Diabetes Care, 26*(7), 1961–1966.

Gruzelier, J., Seymour., K., Wilson, L., Jolley, A. & Hirsch, S. (1988). Impairments on Neuropsychologic Tests of Temporohippocampal and Frontohippocampal Functions and Word Fluency in Remitting Schizophrenia and Affective-Disorders. *Archives of General Psychiatry, 45*, 623–629.

Hall, J. L., Gonder-Frederick, L. A., Chewning, W. W., Silveira, J., & Gold, P. E. (1989). Glucose enhancement of performance of memory tests in young and aged humans. *Neuropsychologia, 27*(9), 1129–1138.

Hope, C., Seiss, E., Dean, P. J. A., Williams, K. E. M., & Sterr, A. (2013). Consumption of glucose drinks slows sensorimotor processing: Double-blind placebo-controlled studies with the Eriksen flanker task. *Frontiers in Human Neuroscience, 7*(October), 1–10.

Hoyer, S. (2003). Memory function and brain glucose metabolism. *Pharmacopsychiatry, 36*(Suppl 1), S62–S67.

Hoyland, A., Lawton, C. L., & Dye, L. (2008). Acute effects of macronutrient manipulations on cognitive test performance in healthy young adults: A systematic research review. *Neuroscience and Biobehavioral Reviews, 32*(1), 72–85.

Jacoby, L. L. (1991). A process dissociation framework: Separating automatic from intentional uses of memory. *Journal of Memory and Language, 30*(5), 513–541.

Kaufman, F. R., Epport, K., Engilman, R., & Halvorson, M. (1999). Neurocognitive functioning in children diagnosed with diabetes before age 10 years. *Journal of Diabetes and its Complications, 13*(1), 31–38.

Kennedy, C., & Sokokloff, L. (1957). An adaptation of the nitrous oxide method to the study of the cerebral circulation in children; Normal values for cerebral blood flow nd cerebral metabolic rate in childhood. *The Journal of Clinical Investigation, 36*, 1130–1137.

Kennedy, D. O., & Scholey, A. B. (2000). Glucose administration, heart rate and cognitive performance: Effects of increasing mental effort. *Psychopharmacology, 149*(1), 63–71.

Korol, D. L., & Gold, P. E. (1998). Glucose, memory, and aging. *American Journal of Clinical Nutrition, 67*, 764–771.

Krebs-Kraft, D. L., & Parent, M. B. (2008). Hippocampal infusions of glucose reverse memory deficits produced by co-infusions of a GABA receptor agonist. *Neurobiology of Learning and Memory, 89*(2), 142–152.

Lamport, D. J., Lawton, C. L., Mansfield, M. W., & Dye, L. (2009). Impairments in glucose tolerance can have a negative impact on cognitive function: A systematic research review. *Neuroscience and Biobehavioral Reviews, 33*(3), 394–413.

Languren, G., Montiel, T., Julio-Amilpas, A., & Massieu, L. (2013). Neuronal damage and cognitive impairment associated with hypoglycemia: An integrated view. *Neurochemistry International, 63*(4), 331–343.

Lieberman, H. R. (2003). Nutrition, brain function and cognitive performance. *Appetite, 40*(3), 245–254.

Luengo-Fernandez, R., Leal, J., & Gray, A. (2010). *Dementia 2010—The economic burden of dementia and associated research finding in the United Kingdom*. Alzheimer's Research Trust.

Macpherson, H., Roberstson, B., Sünram-Lea, S., Stough, C., Kennedy, D., & Scholey, A. (2015). Glucose administration and cognitive function: Differential effects of age and effort during a dual task paradigm in younger and older adults. *Psychopharmacology, 232*(6), 1135–1142.

Magistretti, P. J. (2008). Brain energy metabolism. In L. R. Squire, F. E. Bloom, N. C. Spitzer, S. du Lac, A. Ghosh, & D. Berg (Eds.), *Fundamental neuroscience* (pp. 271–293). Academic Press.

Manning, C. A., Ragozzino, M. E., & Gold, P. E. (1993). Glucose enhancement of memory in patients with probable senile dementia of the Alzheimer's type. *Neurobiology of Aging, 14*(6), 523–528.

Manning, C. A., Stone, W. S., Korol, D. L., & Gold, P. E. (1998). Glucose enhancement of 24-h memory retrieval in healthy elderly humans. *Behavioural Brain Research, 93*(1–2), 71–76.

Maric, N., Doknic, M., Damjanovic, A., Pekic, S., Jasovic-Gasic, M., & Popovic, V. (2008). Glucoregulation in normal weight schizophrenia patients treated by first generation antipsychotics. *Srpski Arhiv za Celokupno Lekarstvo, 136*(3–4), 110–115.

Martins, I. J., Hone, E., Foster, J. K., Sünram-Lea, S. I., Gnjec, A., Fuller, S. J., … Martins, R. N. (2006). Apolipoprotein E, cholesterol metabolism, diabetes, and the convergence of risk factors for Alzheimer's disease and cardiovascular disease. *Molecular Psychiatry, 11*(8), 721–736.

McCarthy, A. M., & Elmendorf, J. S. (2007). GLUT4's itinerary in health & disease. *Indian Journal of Medical Research, 125*(3), 373–388.

McNay, E. C. (2005). The impact of recurrent hypoglycemia on cognitive function in aging. *Neurobiology of Aging, 26*, 76–79.

McNay, E. C., & Gold, P. E. (2001). Age-related differences in hippocampal extracellular fluid glucose concentration during behavioral testing and following systemic glucose administration. *Journals of Gerontology, Series A: Biological Sciences and Medical Sciences, 56*(2), B66–B71.

McNay, E. C., McCarty, R. C., & Gold, P. E. (2001). Fluctuations in brain glucose concentration during behavioral testing: Dissociations between brain areas and between brain and blood. *Neurobiology of Learning and Memory, 75*(3), 325–337.

Meikle, A., Riby, L. M., & Stollery, B. (2004). The impact of glucose ingestion and gluco-regulatory control on cognitive performance: A comparison of younger and middle aged adults. *Human Psychopharmacology, 19*(8), 523–535.
Meikle, A., Riby, L. M., & Stollery, B. (2005). Memory processing and the glucose facilitation effect: The effects of stimulus difficulty and memory load. *Nutritional Neuroscience, 8*(4), 227–232.
Messier, C. (1998). The absence of effect of glucose on memory is associated with low susceptibility to the amnestic effects of scopolamine in a strain of mice. *Behavioural Brain Research, 96*(1–2), 47–57.
Messier, C. (2003). Diabetes, Alzheimer's disease and apolipoprotein genotype. *Experimental Gerontology, 38*, 941–946.
Messier, C. (2004). Glucose improvement of memory: A review. *European Journal of Pharmacology, 490*(1–3), 33–57.
Messier, C., Awad-Shimoon, N., Gagnon, M., Desrochers, A., & Tsiakas, M. (2011). Glucose regulation is associated with cognitive performance in young nondiabetic adults. *Behavioural Brain Research, 222*(1), 81–88.
Messier, C., Desrochers, A., & Gagnon, M. (1999). Effect of glucose, glucose regulation, and word imagery value on human memory. *Behavioral Neuroscience, 113*(3), 431–438.
Messier, C., Gagnon, M., & Knott, V. (1997). Effect of glucose and peripheral glucose regulation on memory in the elderly. *Neurobiology of Aging, 18*(3), 297–304.
Messier, C., Pierre, J., Desrochers, A., & Gravel, M. (1998). Dose-dependent action of glucose on memory processes in women: Effect on serial position and recall priority. *Cognitive Brain Research, 7*(2), 221–233.
Messier, C., Tsiakas, M., Gagnon, M., & Desrochers, A. (2010). Effect of age and glucoregulation on cognitive performance. *Journal of Clinical and Experimental Neuropsychology, 32*(8), 809–821.
Metzger, M. M. (2000). Glucose enhancement of a facial recognition task in young adults. *Physiology and Behavior, 68*(4), 549–553.
Metzger, M. M., & Flint, R. W. (2003). Glucose enhancement of face recognition is unaffected by alterations of face features. *Neurobiology of Learning and Memory, 80*(2), 172–175.
Muraven, M., Tice, D. M., & Baumeister, R. F. (1998). Self-control as a limited resource: Regulatory depletion patterns. *Journal of Personality and Social Psychology, 74*(3), 774–789.
Naguib, J. M., Kulinskaya, E., Lomax, C. L., & Garralda, M. (2009). Neurocognitive performance in children with type 1 diabetes—A meta-analysis. *Journal of Pediatric Psychology, 34*(3), 271–282.

Newcomer, J. W., Selke, G., Melson, A. K., Hershey, T., Craft, S., Richards, K., & Alderson, A. L. (1999). Decreased memory performance in healthy humans induced by stress- level cortisol treatment. *Archives of General Psychiatry, 56*(6), 527–533.

Northam, E. A., & Lin, A. (2010). Hypoglycaemia in childhood onset type 1 diabetes-part villain, but not the only one. *Pediatric Diabetes, 11*(2), 134–141.

Owen, L., Finnegan, Y., Hu, H., Scholey, A. B., & Sünram-Lea, S. I. (2010). Glucose effects on long-term memory performance: Duration and domain specificity. *Psychopharmacology, 211*(2), 131–140.

Owen, L., Scholey, A., Finnegan, Y., & Sünram-Lea, S. I. (2013). Response variability to glucose facilitation of cognitive enhancement. *The British Journal of Nutrition, 110*(10), 1873–1884.

Owen, L., Scholey, A. B., Finnegan, Y., Hu, H., & Sünram-Lea, S. I. (2012). The effect of glucose dose and fasting interval on cognitive function: A double-blind, placebo-controlled, six-way crossover study. *Psychopharmacology, 220*(3), 577–589.

Owens, D. S., Parker, P. Y., & Benton, D. (1997). Blood glucose and subjective energy following cognitive demand. *Physiology and Behavior, 62*(3), 471–478.

Parent, M. B., Krebs-Kraft, D. L., Ryan, J. P., Wilson, J. S., Harenski, C., & Hamann, S. (2011). Glucose administration enhances fMRI brain activation and connectivity related to episodic memory encoding for neutral and emotional stimuli. *Neuropsychologia, 49*(5), 1052–1066.

Parsons, M. W., & Gold, P. E. (1992). Glucose enhancement of memory in elderly humans: An inverted-U dose-response curve. *Neurobiology of Aging, 13*(3), 401–404.

Ragozzino, M. E., & Gold, P. E. (1995). Glucose injections into the medial septum reverse the effects of intraseptal morphine infusions on hippocampal acetylcholine output and memory. *Neuroscience, 68*(4), 981–988.

Ragozzino, M. E., Pal, S. N., Unick, K., Stefani, M. R., & Gold, P. E. (1998). Modulation of hippocampal acetylcholine release and spontaneous alternation scores by intrahippocampal glucose injections. *The Journal of Neuroscience: The Official Journal of the Society for Neuroscience, 18*(4), 1595–1601.

Reger, M. A., Watson, G. S., Frey, W. H., Baker, L. D., Cholerton, B., Keeling, M. L., … Craft, S. (2006). Effects of intranasal insulin on cognition in memory-impaired older adults: Modulation by APOE genotype. *Neurobiology of Aging, 27*(3), 451–458.

Reger, M. A., Watson, G. S., Green, P. S., Baker, L. D., Cholerton, B., Fishel, M. A., … Craft, S. (2008). Intranasal insulin administration dose-dependently modulates verbal memory and plasma amyloid-beta in memory-impaired older adults. *Journal of Alzheimer's Disease: JAD, 13*(3), 323–331.

Riby, L. M. (2004). The impact of age and task domain on cognitive performance: A meta-analytic review of the glucose facilitation effect. *Brain Impairment, 5*(2), 145–165.

Riby, L. M., Lai Teik Ong, D., Azmie, N. B. M., Ooi, E. L., Regina, C., Yeo, E. K. W., … Aquili, L. (2017). Impulsiveness, postprandial blood glucose, and glucoregulation affect measures of behavioral flexibility. *Nutrition Research, 48*, 65–75.

Riby, L. M., Marriott, A., Bullock, R., Hancock, J., Smallwood, J., & McLaughlin, J. (2009). The effects of glucose ingestion and glucose regulation on memory performance in older adults with mild cognitive impairment. *European Journal of Clinical Nutrition, 63*(4), 566–571.

Riby, L. M., McMurtrie, H., Smallwood, J., Ballantyne, C., Meikle, A., & Smith, E. (2006). The facilitative effects of glucose ingestion on memory retrieval in younger and older adults: Is task difficulty or task domain critical? *British Journal of Nutrition, 95*(2), 414.

Riby, L. M., Meikle, A., & Glover, C. (2004). The effects of age, glucose ingestion and gluco-regulatory control on episodic memory. *Age and Ageing, 33*(5), 483–487.

Riby, L. M., Sünram-Lea, S. I., Graham, C., Foster, J. K., Cooper, T., Moodie, C., & Gunn, V. P. (2008). P3b versus P3a: An event-related potential investigation of the glucose facilitation effect. *Journal of Psychopharmacology, 22*(5), 486–492.

Ruis, C., van den Donk, M., Biessels, G. J., Kapelle, L. J., Gorter, K., & Rutten, G. E. H. (2009). Cognition in the early stage of type 2. *Diabetes Care, 32*(7), 3–7.

Ryan, C. M. (2006). Diabetes and brain damage: More (or less) than meets the eye? *Diabetologia, 49*(10), 2229–2233.

Ryan, C. M., & Geckle, M. (2000). Why is learning and memory dysfunction in Type 2 diabetes limited to older adults? *Diabetes/Metabolism Research and Reviews, 16*(June), 308–315.

Scholey, A., MacPherson, H., Sünram-Lea, S., Elliott, J., Stough, C., & Kennedy, D. (2013). Glucose enhancement of recognition memory: Differential effects on effortful processing but not aspects of "remember-know" responses. *Neuropharmacology, 64*(July), 544–549.

Scholey, A. B., Harper, S., & Kennedy, D. O. (2001). Cognitive demand and blood glucose. *Physiology and Behavior, 73*(4), 585–592.

Scholey, A. B., Laing, S., & Kennedy, D. O. (2006). Blood glucose changes and memory: Effects of manipulating emotionality and mental effort. *Biological Psychology, 71*(1), 12–19.

Scholey, A. B., Sünram-Lea, S. I., Greer, J., Elliott, J., & Kennedy, D. O. (2009a). Glucose administration prior to a divided attention task improves tracking performance but not word recognition: Evidence against differential memory enhancement? *Psychopharmacology, 202*(1–3), 549–558.

Scholey, A. B., Sünram-Lea, S. I., Greer, J., Elliott, J., & Kennedy, D. O. (2009b). Glucose enhancement of memory depends on initial thirst. *Appetite, 53*(3), 426–429.

Schultz, S. K., Arndt, S., Ho, B. C., Oliver, S. E., & Andreasen, N. C. (1999). Impaired glucose tolerance and abnormal movements in patients with schizophrenia. *The American Journal of Psychiatry, 156*(4), 640–642.

Seidman, L. J., Talbot, N. L., Kalinowski, A. G., Mccarley, R. W., Faraone, S. V., Kremen, W. S., Pepple, J. R. & Tsuang, M. T. (1991). Neuropsychological Probes of Fronto-Limbic System Dysfunction in Schizophrenia—Olfactory Identification and Wisconsin Card Sorting Performance. *Schizophrenia Research, 6*, 55–65.

Smith, M. A., & Foster, J. K. (2008). Glucoregulatory and order effects on verbal episodic memory in healthy adolescents after oral glucose administration. *Biological Psychology, 79*(2), 209–215.

Smith, M. A., Riby, L. M., Sünram-Lea, S. I., van Eekelen, J. A. M., & Foster, J. K. (2009). Glucose modulates event-related potential components of recollection and familiarity in healthy adolescents. *Psychopharmacology, 205*(1), 11–20.

Smith, M. A., Riby, L. M., van Eekelen, J. A. M., & Foster, J. K. (2011a). Glucose enhancement of human memory: A comprehensive research review of the glucose memory facilitation effect. *Neuroscience and Biobehavioral Reviews, 35*(3), 770–783.

Smith, M. A., Hii, H. L., Foster, J. K., & Van Eekelen, J. A. M. (2011b). Glucose enhancement of memory is modulated by trait anxiety in healthy adolescent males. *Journal of Psychopharmacology, 25*(1), 60–70.

Stefani, M. R., Nicholson, G. M., & Gold, P. E. (1999). ATP-sensitive potassium channel blockade enhances spontaneous alternation performance in the rat: A potential mechanism for glucose-mediated memory enhancement. *Neuroscience, 93*(2), 557–563.

Stone, W. S., & Hsi, X. (2011). Declarative memory deficits and schizophrenia: Problems and prospects. *Neurobiology of Learning and Memory, 96*(4), 544–552.

Stone, W. S., Seidman, L. J., Wojcik, J. D., & Green, A. I. (2003). Glucose effects on cognition in schizophrenia. *Schizophrenia Research, 62*(1–2), 93–103.
Stone, W. S., Thermenos, H. W., Tarbox, S. I., Poldrack, R. A., & Seidman, L. J. (2005). Medial temporal and prefrontal lobe activation during verbal encoding following glucose ingestion in schizophrenia: A pilot fMRI study. *Neurobiology of Learning and Memory, 83*(1), 54–64.
Strachan, M. W., Deary, I. J., Ewing, F. M., & Frier, B. M. (1997). Is type II diabetes associated with an increased risk of cognitive dysfunction? A critical review of published studies. *Diabetes Care, 20*(3), 438–445.
Sünram-Lea, S. I., Dewhurst, S. A., & Foster, J. K. (2008). The effect of glucose administration on the recollection and familiarity components of recognition memory. *Biological Psychology, 77*(1), 69–75.
Sünram-Lea, S. I., Foster, J. K., Durlach, P., & Perez, C. (2001). Glucose facilitation of cognitive performance in healthy young adults: Examination of the influence of fast-duration, time of day and pre-consumption plasma glucose levels. *Psychopharmacology, 157*(1), 46–54.
Sünram-Lea, S. I., Foster, J. K., Durlach, P., & Perez, C. (2002a). Investigation into the significance of task difficulty and divided allocation of resources on the glucose memory facilitation effect. *Psychopharmacology, 160*(4), 387–397.
Sünram-Lea, S. I., Foster, J. K., Durlach, P., & Perez, C. (2002b). The effect of retrograde and anterograde glucose administration on memory performance in healthy young adults. *Behavioural Brain Research, 134*(1–2), 505–516.
Swanson, R. (1992). Physiologic coupling of glial glycogen metabolism to neuronal activity in brain. *Canadian Journal of Physiology and Pharmacology, 70*, S138–S144.
Watson, G. S., & Craft, S. (2004). Modulation of memory by insulin and glucose: Neuropsychological observations in Alzheimer's disease. *European Journal of Pharmacology, 490*(1–3), 97–113.
White, N. N. (1991). Peripheral and central memory—Enhancing actions of glucose. *Peripheral Signalling of the Brain, 6*, 421–441.
Wright, R. S., Levy, S. A. T., Katzel, L. I., Rosenberger, W. F., Manukyan, Z., Whitfield, K. E., & Waldstein, S. R. (2015). Fasting glucose and glucose tolerance as potential predictors of neurocognitive function among nondiabetic older adults. *Journal of Clinical and Experimental Neuropsychology, 37*(1), 49–60.
Zacks, R. T., Hasher, L., & Li, K. Z. H. (2000). Human memory. In *The Handbook of Aging and Cognition* (pp. 293–357). Lawrence Erlbaum Associates.

7

A Familiar Landscape in the Brave New World: Ethics of Cognitive Enhancement Introduction

Vince Cakic

In 1937 and responding to media reports that college students in the Midwest had been using amphetamine while cramming for their finals, an editorial in *JAMA* voiced concerns over the growing misuse of what an article in *Time* magazine would go to describe as a "powerful but poisonous brain stimulant" ("Pep-Pill Poisoning", 1937). The students, it was claimed, had been prompted by "exaggerations in newspaper accounts" (p. 1973) likening amphetamine to "high octane number gasoline" for the brain that allowed it to "hit on all cylinders" (New York Times, 1937):

> It is chiefly the ignorant who try such self-medication, not realizing that a drug can never substitute for knowledge or intellect. The drug is too new to pharmacology and experimental medicine to warrant any prediction as to possible permanent harm that may result from its continued misuse. (Editorial, 1937, p. 1974)

V. Cakic (✉)
University of Sydney, Sydney, NSW, Australia

© The Author(s) 2021
M. Hall et al. (eds.), *Chemically Modified Minds*,
https://doi.org/10.1007/978-981-15-6771-1_7

Cognitive enhancement (CE) has emerged as some sort of cutting-edge conundrum within neuroethics and a sign of the times for a species seeking to apply its technology inward in a hubristic attempt to transcend their own biological limitations. Yet for all the warnings of the moral hazards that lie ahead (Kass, 2003), the future we have before us seems a little less like a Brave New World, and reads more like a Philip K. Dick paperback that never quite made it to print; an anachronistic dystopia where cosmetic neurologists will surely run a brisk trade rigging amphetamine-grade bootstrap lifters into anyone game enough to tinker with the pistons of their internal combustion engines.

In 2011, the "baby boomers" officially commenced retirement and, with them, a population exceeding half a billion susceptible to age-related cognitive decline. Yet for what represents a clear commercial incentive to develop effective CE, one cannot help but wonder if in all the excitement over the coming era of cosmetic neurology, did everybody forget to invite any of the smart drugs to the party?

This chapter will provide an overview of the landscape that has come to shape the discourse on CE. The use of psychostimulants by college students is not a new phenomenon, and much of the controversy surrounding CE has been shaped by last century's pharmacopoeia. I argue that the focus on these agents as CE has led to a definitional drift, and the discussion should be re-oriented towards intelligence enhancement (IE), which although also a phantom debate (see Quednow, 2010) at least presents more fertile grounds for discourse in neuroethics.

Re-defining Cognitive Enhancement

Before considering the main themes that have emerged from the literature on CE, I will note some peculiarities that have come to shape it.

Cognitive Enhancers Should Enhance Cognitive Ability

If, as E. G. Boring (1923) famously wrote "intelligence is what the tests test" (p. 35), then perhaps an equally banal definition is needed for

CE. Cognitive enhancement should seek as it ends to increase the individual's cognitive ability. And if this is to be the case, then CE should aspire as its penultimate goal to modify what cognitive ability has traditionally been understood to mean, the general factor of intelligence, as defined by psychometric *g* (see Jensen, 1998). Cognitive enhancement ought to improve whatever the intelligence tests are testing, an endeavour that working memory training (Jaeggi, Buschkuehl, Jonides, & Perrig, 2008; but see Melby-Lervåg, Redick, & Hulme, 2016) appears closer to realising than any pharmacological intervention currently can or is likely to in the near future.

Psychostimulants Are a Snake-Oil

Although Parens (1998) correctly notes that with each attempt to define enhancement, the term becomes more and more meaningless, a more pervasive problem in the literature on CE is what can only be characterised as *definitional drift*. Getting good grades may be a hobby of those with greater cognitive ability, but agents used to improve academic performance are not necessarily CE. The college students' use of amphetamine to cram for finals has now come to represent a modern-day snake-oil, for example, use of psychostimulants is viewed as more necessary for success in the academic domain than the use of steroids in sports (Dodge, Williams, Marzell, & Turrisi, 2012). This is in spite of the finding that these agents do not appear to confer any benefit in those without a diagnosis of attention deficit hyperactivity disorder (ADHD) (Arria et al., 2017) and only modest improvements in academic performance among those who do (Baweja, Mattison, & Waxmonsky, 2015).

Psychostimulants May Act as Motivation Enhancers

While there is little evidence that dopaminergic psychostimulants have efficacy with in those with normal baseline cognitive performance (Ilieva, Boland, & Farah, 2013), for lack of a better candidate, these drugs are the most commonly considered agents in the literature (Quednow, 2010).

Equally pervasive is the characterisation of these agents as CE when their effects are more reminiscent of motivation enhancers (Volkow et al., 2004). Early clinical work examined amphetamine as a pharmacotherapy for anhedonia and its efficacy in attenuating boredom (Barmack, 1938).

Only a handful entertain the possibility that the dopaminergic psychostimulants act as motivation enhancers (Ilieva & Farah, 2013; Quednow, 2010; Vrecko, 2013), while further still, others (e.g. Lavazza, 2016) opine that agents acting on the mesolimbic pathways exert their effects on anything *except* motivation.

Dangers

In the film *Limitless* (2011) the film's protagonist is able to transcend his own mediocrity care of NZT-48, a fictional drug that imbues him with superintelligence. However, as his use escalates and his supply is cut, he is soon acquainted with the harsh reality that there is no such thing as a psychopharmacological free lunch. The film accurately captures the current *Zeitgeist* surrounding CE, because whether accident or design, the dopaminergic stimulants—methylphenidate (Ritalin), amphetamine (Adderall), methamphetamine (Desoxyn)—have become the *de facto* standard-bearers of what CE represents.

Not only are the risks of CE the most commonly cited concern by the public (Schelle, Faulmüler, Caviola, & Hewstone, 2014) but also shape the acceptability of CE in both the public (Scheske & Schnall, 2012) and the physicians who would conceivably be the ones to prescribe them. And while there is no reason that we ought to assume any CE could ever be perfectly safe, one cannot help but wonder whether the use of an agent that comes with the risk of addiction, neurotoxicity, and psychosis sounds less like CE and a little more like a Faustian bargain.

However, the dangers of CE alone cannot wholly account for the opposition towards CE when any number of activities carry with them serious risk of injury or death, yet they appear bereft of the same moral judgements (e.g. Nutt, 2009). For example, Ivy League colleges offer preferences to athlete applicants equivalent to an additional 160 SAT points (Espenshade & Chung, 2005) and those who play intercollegiate football are subjected to 1000 head impacts per season (Crisco et al., 2010). One wonders why the student-athlete whose acceptance into an Ivy League college does not evince the same moral judgements in the public (Ott, Bozeman, & Taggart, 2018) when it carries with it the very real risk of chronic traumatic encephalopathy (Mez et al., 2017).

With the prominence of psychostimulants in shaping opinions on CE, much of the debate regarding its safety and the moral judgements comes off the back of almost half a century of drug prohibition (Coomber, 2014). The language of drug prohibition seems to have embedded itself into the discourse on CE as well. Thus, *soft enhancers*—exercise, caffeine, herbs—are now distinguished from the *hard* enhancers (Liakoni, Schaub, Maier, Glauser, & Liechti, 2015). That the delineation between soft and hard enhancement is not defined by efficacy (cf. Neale, Camfield, Reay, Stough, & Scholey, 2013), but legal status as prescription drugs is no coincidence.

Treatment Versus Enhancement

The normative perspective on CE has centred itself around the belief that increasing the cognitive ability of healthy individuals is, to use the vernacular of the President's Council of Bioethics (2003), "beyond therapy". This conception of enhancement is premised on the assumption that there is a clear distinction between treatment and enhancement (Partridge, Lucke, & Hall, 2014) and as practitioners of the use of a particular drug for a particular purpose. I believe that the fundamental shortcoming from such a discourse is that while authors acknowledge the arbitrariness of this distinction, they still attempt to frame CE under this paradigm.

Proper Goals of Medicine

The bioethics principle of beneficence obliges that a therapeutic intervention can be justified only if its expected benefits outweigh any of its risks. This is particularly salient in the case of CE, where the absence of any recognised disease entity might render the existence of any non-zero risk unjustifiable. This is a sentiment that comports with physician attitudes regarding enhancement (Banjo, Nadler, & Reiner, 2010). While the principle of beneficence should always guide any clinical decision, the very real risk that is associated with cosmetic surgery (Chatterjee, 2007) suggests that even if they cannot be clinically justified, nontherapeutic interventions could very well be subsumed into the everyday practice of conventional medicine.

The fundamental shortcoming of this approach, however, is that the purpose of a ligand is not to treat disease (Bostrom, 2008), nor has it ever been. It is to bind to a receptor and instantiate any desired biological effect, the ability to arrest the disease process being one of many end-points either wholly embraced or accepted by society. In the mid-1800s, Snow popularised the use of chloroform in childbirth, a practice that some resisted because it "does not appear to possess any peculiar property, rendering it useful in midwifery, except by removing the sensation of pain" (Merriman, 1848, p. 21). And so it began—the medicalisation of unpleasant quale through obstetric anaesthesia.

Oral contraceptives allow women to uncouple sexual activity from the procreative process, yet pregnancy is not a disease. Ethyl alcohol is only indicated for ethylene glycol poisoning, but we seem to prefer it more as an inebriant. Androgenic alopecia may have its own ICD-10 designation and finasteride may be an FDA-approved treatment, but to claim that male pattern baldness is a disease that ought to be subsumed under the rubric of therapy seems to suggest that medicalisation and medicalese (Young, Norman, & Humphreys, 2008) are all that is required for enhancement's moral comb-over into "proper medicine".

Intelligence Enhancement as Therapy

The current conception of CE has been shaped by the pharmacopoeia presently available to us. It has construed better academic performance, an outcome typically seen in those with higher cognitive ability as the penultimate goal of CE, in the process creating a caricature of what the true aspirations of CE ought to be. I have previously presented a definition of CE as an agent that increases cognitive ability, with the view that any meaningful CE will result in intelligence enhancement (IE).[1]

Although the therapeutic potential of IE has not previously been expounded, the view that it could be therapeutic is not entirely new either (Bostrom & Sandberg, 2009; Dunlop & Savulescu, 2014). For example, Savulescu (2006) correctly argues that IE in a child would likely result in a good life, but he does not consider that the good life he speaks of falls well within purview of conventional medicine. Here, I will argue that intelligence enhancement represents both an enhancer and therapy in equal measure.

I. The Ideal Enhancer

Rather than merely improving the achievement outcomes, for example academic performance to which intelligence is correlated, the ideal enhancer is one that would increase *g* by acting on the biological substrate that underpins it. Consider the Flynn Effect, the phenomenon wherein every 15 years the intelligence of the population increases 5 IQ points (Trahan, Stuebing, Fletcher, & Hiscock, 2014). Suppose through one-part psychopharmacology and one-part alchemy, the Flynn Effect over a generation could be transmuted into a pharmacotherapy, and it could instantiate in an individual all the benefits and harms associated with having an IQ 1 SD higher.

To be sure, while I do not hold any optimism that ligand-based enhancement could ever confer this type of effect in the healthy, the magnitude of benefit from our fictional drug is comparable to the loss of 13.5 points of IQ that salt iodisation prevents in maternal iodine deficiency (Zimmermann, 2012). Notwithstanding, as a thought experiment let us suppose that our IE drug could effect a 1 SD increase in IQ along the entire distribution of intelligence and like iodine supplementation, must be consumed during early pregnancy.

---

[1] For conceptual clarity, here I will use intelligence enhancement (IE) to distinguish from the drugs currently subsumed under CE.

II. The Therapeutic Claim

The basis for IE as a therapeutic intervention is a simple one and can be readily subsumed within the extant paradigm: if increasing an individual's IQ by 1 SD would result in a meaningful improvement in health outcomes that fall within the scope of conventional medicine, the use of IE ought not to be couched differently to any other therapeutic claim (Fig. 7.1).

Although a thorough review is beyond our scope, there is a consistent and robust association between IQs measured in childhood or early adulthood and all-cause mortality along the entire range of intelligence (see Calvin, Batty, & Dreary, 2011). Thus, if we could increase an individual's IQ by 1 SD, we can expect to see roughly a 25% reduction in the risk of all-cause death at ages 40 and 70. This relationship has been well replicated (Hart et al., 2003; Wrulich, Stadler, Brunner, Keller, & Martin, 2015) and is supported by meta-analysis (Calvin et al., 2011).

Lower intelligence is also a risk factor for a multiplicity of cause-specific morbidity (Batty, Deary, Schoon, & Gale, 2007) and mortality (Calvin et al., 2017; Christensen, Mortensen, Christensen, & Osler, 2016). Gale et al. (2010), for example, report that a 1 SD reduction in IQ increased the risk of hospitalisation for various psychiatric illnesses including schizophrenia (HR = 1.66), mood (HR = 1.49), and alcohol (HR = 1.72). Similar links between pre-morbid IQ and psychiatric illness are reported elsewhere (Urfer-Parnas, Lykke Mortensen, Sæbye, & Parnas, 2010; Zammit et al., 2004). Lower IQ also predicts hospitalisation for unintentional injury (HR = 1.15; Whitley et al., 2010b), assault (HR = 1.51; Whitley et al., 2010a), mortality from homicide (Batty, Deary, Tengstrom, & Rasmussen, 2008), and motor vehicle accident (O'Toole, 1990).

There are several important things to note here. First, the relationship between lower intelligence and poorer health outcomes seems to exist along the entire distribution of cognitive ability. Moreover, the relationship is non-linear and a 1 SD increase in intelligence would be most impactful to the health outcomes of those at the lower end of cognitive ability (Wrulich et al., 2015). Just as importantly, social inequalities do not seem to be driving the relationship between intelligence and health

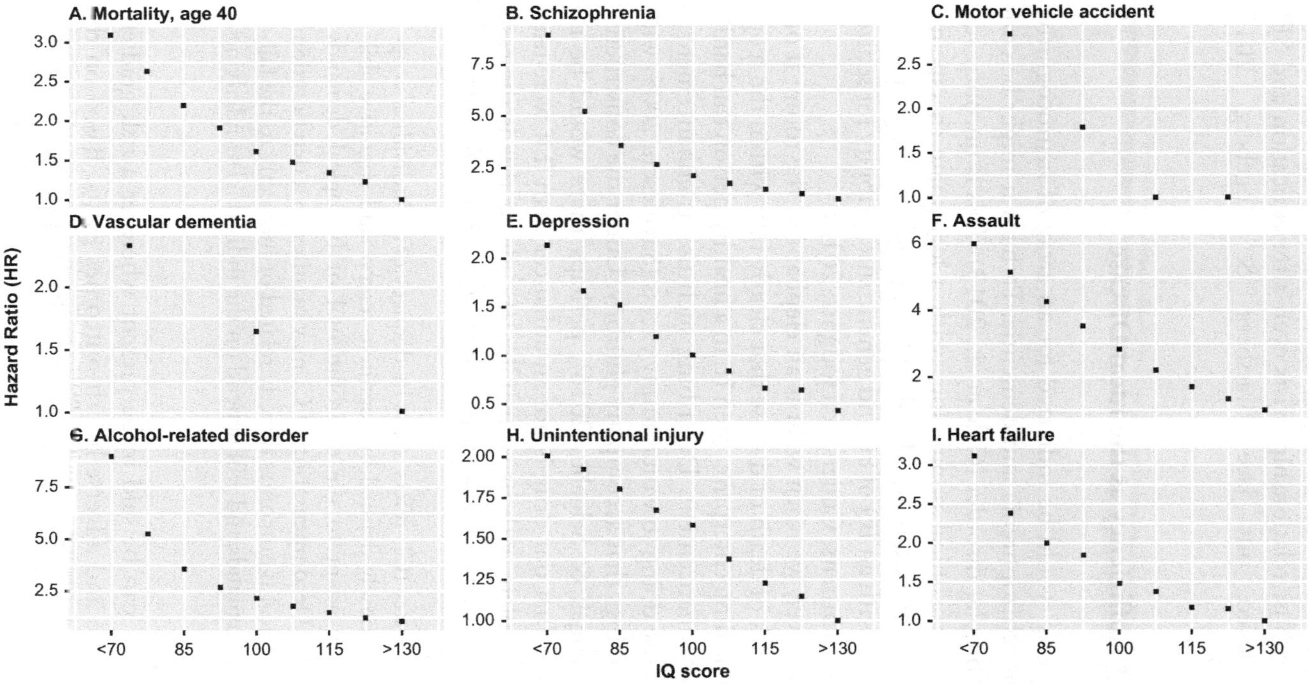

Fig. 7.1 Risk of various health outcomes per 1 SD reduction in IQ from comparison group in early adulthood. (**a**) All-cause mortality by middle age (Batty et al., 2009). (**b**) Schizophrenia hospitalisation (Gale, Batty, Tynelius, Deary, & Rasmussen, 2010). (**c**) Motor vehicle accident mortality (O'Toole, 1990). (**d**) Vascular dementia diagnosis (Osler, Christensen, Garde, Mortensen, & Christensen, 2017). (**e**) Depression hospitalisation (Lager, Melin, Hemmingsson, & Wallin, 2017). (**f**) Assault hospitalisation (Whitley et al., 2010a). (**g**) Alcohol-related disorder hospitalisation (Gale et al., 2010). (**h**) Unintentional injury hospitalisation (Whitley et al., 2010b). (**i**) Heart failure hospitalisation (Lindgren et al., 2018)

inequalities as a robust relationship exists even after education and SES have been factored out (see Bratsberg & Rogeberg, 2017; Gottfredson, 2004).

III. Cognitive Insufficiency as the Ideal Disease

No doubt, our drug would be indicated for an array of neurocognitive disorders such as schizophrenia and dementia. Perhaps it might be contraindicated for bipolar disorder (Gale et al., 2013). This is uncontroversial. Eventually, non-clinical populations with a legitimate therapeutic need would emerge. For example, IE might be appropriate prophylaxis in an adolescent male with an IQ of 80 and genetic susceptibility for schizophrenia (Dickson, Laurens, Cullen, & Hodgins, 2012).

Eventually, however, we reach what is an elephant in the room: if our drug reduces the risk of death or disease that clearly falls within the scope of medicine as a healing enterprise, then IE in a healthy person can hardly be thought of as enhancement. Describing such use as enhancement is misnomeric because doing so would imply excess (Savulescu, 2005) with all the moral trappings that accompany gluttony. Under the constructivist paradigm (Gyngell & Selgelid, 2016), perhaps there would be no greater affirmation that our drug is not an enhancer than if it were used to treat a disease replete with its own ICD-10 code. If this is the case, then we ought to construct a new entity called *cognitive insufficiency* (CI):

> Cognitive insufficiency (CI) refers to the inadequacy of an individual's cognitive resources—loosely defined by cognitive reserve and cognitive ability—to adequately adapt to the environmental or biological demands placed upon them. Decompensation of CI may manifest as a deterioration in any number of health outcomes directly related or indirectly to the failure to meet these demands and will depend on the individual's underlying diathesis.

Conceptually, CI can be said to be present wherever a risk gradient between intelligence and adverse health outcomes exists. Based on the findings of longitudinal studies, CI appears to be present to varying degrees in any individual with an IQ of below 130, or approximately 97.5% of the population.[2] Nosologically, CI most closely resembles essential hypertension because a cut-off IQ score of 130 is just as arbitrary as the designation of 140/90 mmHg for hypertension.

IV. Treatment and Enhancement Are Not Mutually Exclusive

Ostensibly, IE offers us the possibility to intervene in an array of health outcomes subsumed by proper medicine. Yet we can just as equally expect that our treatment for CI would also improve our standing in other domains of our life. While these positional enhancements could very well overshadow the therapeutic benefits of IE, they need not necessarily render them any less legitimate (Savulescu, 2005). Consider the following:

> Jan is a 40-year-old female with a history of refractory depression. Unhappy with the presence of facial wrinkles, she asks her physician for injections of Botulinum toxin A (Botox) into her glabellar muscles. Although he is not a cosmetologist, he nevertheless agrees.

Why might Jan's physician view her request as therapeutic when it has all the hallmarks of enhancement? Perhaps we could reason that the reduction of wrinkles might improve her self-esteem, and represents a type of therapy. Although plausible, this is spurious. Instead, by paralysing the muscles in her face, Jan is unable to frown. Afferents projecting from the face to limbic pathways have a modulating effect on her mood (Finzi & Rosenthal, 2016). The claim that Botox is an antidepressant is perfectly cogent (Magid et al., 2015).[3]

This example has relevance to IE. An individual's primary motivations may be cosmetic, but this would not obviate IE as a legitimate treatment if it confers meaningful health outcomes no different in degree and type to other therapeutic agents. Any improvement in positional standing, for

---

[2] However, in their study of gifted children whose IQ ranged from 135 to 163, Martin and Kubzansky (2005) report a 32% decreased risk of mortality per 1 SD increase in IQ. This suggests that an intelligence-mortality gradient exists even at the highest end of the IQ distribution.

[3] Although Botox is not indicated for depression, it nevertheless displays a robust antidepressant effect (Cohen's $d = 1.07$; Magid et al., 2015).

example academic achievement, ought to be viewed as a side-effect of the intelligence enhancing treatment, albeit a desired one.[4]

While Schermer and Bolt (2011) are perfectly correct in characterising the prospect of IE as nothing more than a transhumanist fantasy, I do not share their view that indulging the hypothetical scenario I have outlined can offer little to the ethical discourse. On the contrary, IE thoroughly undermines any attempt to delineate treatment from enhancement because it represents both a positional and non-positional good in equal measure. When life is effectively the outcome of "one long mental test battery" (Gottfredson, 2003, p. 243), we are confronted with an agent whose biological substrate leads to just as many health outcomes of interest to the physician as it does social inequalities of moral consequence.

Nor would the view that CI amounts to disease mongering be entirely relevant either (Moynihan, Heath, & Henry, 2002). Like essential hypertension (see Pickering, 1968; Rothstein, 2003), the treatment of cognitive insufficiency would represent medicalisation at its finest. One hardly imagines a physician giving pause to whether CI is a real disease entity any more than they do hypertension. After enough generations, any controversy about CI would become just another footnote in the annals of medicine.

Fairness

Much of the discourse around the fairness of CE is focussed on its use in the context of academia. As the paradigmatic example of how ligand-based enhancement amounts to cheating in competitive domains, this generally invites comparisons to the use of doping in sport (Cakic, 2009; Rose, 2005). If an advantage is any opportunity, intervention or strategy offering *some* improvement in performance, and unfairness can be understood in terms of its uneven distribution in society, then the current use of CE can certainly be viewed as an *unfair advantage*. Yet life seems to abound with a very many unfair advantages that unless we are suffering

[4] This should be contrasted with the treatment of ADHD; the pathology itself is a functional impairment in one's positional standing.

from parallax error, it is difficult to believe that a level playing field could ever exist.

Nature of Activity

Any moral claim that CE is unfair is most compelling in zero-sum competitions, where any gain derived from CE obliges a commensurate loss to another. Generally, whenever moral distinctions have been made regarding acceptable use of CE, they have done so along these lines (e.g. Goodman, 2010; Santoni de Sio, Robichaud, & Vincent, 2014). By and large public opinion mirrors these sentiments (Partridge et al., 2014), and while high-stakes college entrance exams are certainly zero-sum, an empirical focus on this population seems all but lacking (but see Teter, DiRaimo, West, Schepis, & McCabe, 2018).

Instead, the literature has oriented itself towards the phenomenon of psychostimulant use by college students. While such use is certainly of relevance to epidemiology (Arria et al., 2017) and college students offer a convenient recruitment pool for investigation of attitudes towards CE (e.g. Scheske & Schnall, 2012), the claim that students utilising CE are deriving an unfair advantage is one that seems to be premised on some assumption that the distribution of college grades is a zero-sum competition. This assumption is difficult to square with the phenomenon of "grade inflation" or "grade compression", which has resulted in a progressive increase in mean GPA scores across institutions over time (Donaldson & Gray, 2012; Johnson, 2003), with the more selective Ivy League colleges appearing to be the worst culprits.

There are several important implications here. Not only is there a tendency for people to see zero-sum competition where in fact there is none (Meegan, 2010; Norton & Sommers, 2011), but such beliefs tend to reduce cooperation (Sirola & Pitesa, 2017) and foment hostility towards perceived rule-breakers (Burleigh, Rubel, & Meegan, 2017; Wellman, Liu, & Wilkins, 2016). It is likely that the public's sentiments towards CE as an unfair advantage are shaped by underlying beliefs in a zero-sum competition. As Hupli (2013) notes of one respondent, "you're not competing against other people you're just competing against yourself… I mean if I get a [grade of] 7 and somebody else gets an 8 it doesn't affect me" (p. 45).

Magnitude of Advantage

If the use of caffeine is deemed morally permissible by the public, but not other forms of CE (e.g. Forlini & Racine, 2012), then perhaps it is a difference of degree of performance improvement that makes CE unfair. Although there has been a focus on how beliefs surrounding the prevalence of CE are likely driving the use of psychostimulants (Schleim & Quednow, 2017), little attention has been given to quantifying the perceived benefits of CE. This shortcoming is two-fold. Not only is there a dearth of evidence to support the efficacy of CE, but any exaggerated perception of their benefit is equally likely with increased sense of unfairness from their use.

Dodge et al. (2012) found that students consider the use of psychostimulants more necessary for academic success than the use of anabolic steroids for athletic success. One cannot help but wonder, if likening CE to "steroids for the brain" and the belief that Herculean academic achievements cannot be accomplished without their use lends itself the belief that CE poses an unfair advantage. The view then that psychostimulants are morally equivalent to caffeine may not necessarily be a claim regarding safety, but that the magnitude of any advantage ought to be considered on par with that of caffeine. Hupli (2013), for example, notes that respondents did not view the use of CE to be unfair, precisely because they perceived their effects to be minimal.

Just how these beliefs have emerged is unclear. In the presence of incomplete information or ambiguity, the tendency to resort to heuristics can result in cognitive biases. Although such biases influencing attitudes have been framed as moral judgements of CE (Caviola, Mannino, Savulescu, & Faulmuller, 2014), it is possible that these cognitive biases are driving beliefs about their efficacy. Thus, while the view that drugs with fewer side-effects could be a moral claim about risk-taking, the very presence of these side-effects may also influence perceptions of greater efficacy (Kramer, Irmak, Block, & Ilyuk, 2012). Likewise, while Caviola et al. (2014) interpret the greater acceptability of herbal enhancers as influencing beliefs regarding safety, the "nature bias" may just as equally be moderate beliefs about lack of efficacy (Lynch & Berry, 2007).

Normative Attitudes Towards Enhancement

In zero-sum competitions, effective CE would certainly confer unfair advantage over non-users. However, it is difficult to conceive how this alone is different to any other advantage that is unequally distributed. Parallels can be seen with the use of erythropoietin (EPO) in sport; while the use of hypoxic air machines is permitted, recombinant human EPO is not (Savulescu, Foddy, & Clayton, 2005). Both modalities will instantiate erythropoiesis, resulting in the desired effect of increasing an athlete's oxygen-carrying capacity, yet only the environmental intervention—many orders of magnitude more expensive than EPO—is bereft of any of the moral trappings of ligand-based performance enhancement (Spriggs, 2005).

The underlying premise appears to be that the outcomes achieved through CE differ in some meaningful way from what is merely a competitive edge. Enhancement, however, might not be viewed as unfair simply because it confers any advantage per se, but because this advantage comes about through the violation of one or more norms. Although just what rule has been broken is unclear, we might speculate that the student who engages in CE cannot do so without also breaking the law and subjecting themselves to some risk of harms—a harm that the non-user perceives to be significant (Arria, Caldeira, Vincent, O'Grady, & Wish, 2008). Thus, while the non-user might appraise CE to confer substantial benefits by way of performance, they are equally unwilling to subject themselves to the risks that such use would oblige. Deriving an advantage from the use of CE, therefore, can only come from the norm-abiding self-restraint of another (Green, 2004).

That CE is less morally acceptable than traditional forms of academic cheating certainly lends itself to this view (Dubljević, Sattler, & Racine, 2014) and while ostensibly there is merit to the claim that CE is unfair, I believe such a claim is ultimately a facile one. To borrow a term from jurisprudence, any unfair advantage that CE confers is the result of *malum prohibitum* violations or actions which are wrong only because they are against the rules (Green, 2004).

Any unfair advantage derived from CE is one that could be rendered moot by making use of CE lawful through medically supervised use of these agents (Greely et al., 2008; Harris, 2011; Schermer, 2008). This can be contrasted with *malum in se* violations—actions that are intrinsically wrongful, irrespective of legal proscriptions. In any society grounded in the belief that success ought to be earned, I can think of no greater *malum in se* violation of the meritocratic ideal than rewards that can be acquired through nepotism or money (see Chen & Tyler, 2001).

I. Legacy Admissions

The admissions preference for the children of alumni, or legacy candidates, is a practice that is almost exclusive to the United States, although there is a case to be had that deriving an advantage alone one's bloodline harkens back to feudal times (Shadowen, Tulante, & Alpern, 2009). Primary legacies, or applicants whose parent attended the institution as an undergraduate, have a robust advantage when applying for the most selective colleges (OR = 14.61; Hurwitz, 2011) while Espenshade and Chung (2005) estimate that the preferential selection weighting that elite universities grant to legacies is the equivalent of an additional 160 SAT points. In the case of the latter, such an advantage confers an increase in rank of roughly three-quarters of a million places.[5]

II. Money for Merit

In my native Australia, an equally favourable admissions pathway is open to those lacking merit. University tuition fees of domestic students are partially subsidised by the Commonwealth government. However, provided that they are willing to forego the subsidy, domestic full-fee students may be accepted into courses five centiles below the admissions

[5] College Board (2017) SAT distribution using the Harvard admission cut-off score of 1540.

cut-off. Perhaps we might agree with Rousseau's (1992/2006) view that wealth is the most basic form of moral inequality, were it not for the fact that this merit can be acquired for as little as $2000 per centile[6] and interest-free, no less. Regardless, there is still something unsettling about what amounts to the acquisition of merit by fiat, when such an arrangement is decidedly skewed in favour of those from a higher SES (Jerrim, Chmielewski, & Parker, 2015).

Inequality

Although it would appear that privilege remains the gold standard "cognitive enhancer", concerns have also been raised that CE will only be accessible to the wealthy few, thereby exacerbating the inequalities that already stratify society. These concerns are well-placed. Today, the college degree has become the new high school diploma and as income inequality widens, so too does intergenerational immobility. Mediating this phenomenon is the growing importance of educational attainment and the incentive for parents to leverage their financial resources to attain any possible advantage they can for their children.

In the United States, smatterings of the type of inequality that can be expected from future CE agents seem to have already emerged. Despite its prevalence being inverse to SES (Froehlich et al., 2007), children from a higher SES have a higher rate of psychostimulant treatment for ADHD (Simoni & Drentea, 2016). Moreover, King, Jennings, and Fletcher (2014) found children from a higher SES are more likely to engage in the selective use of stimulants during the academic year, lending to the view that health inequalities may be leveraged by families of a higher SES a means of transmitting educational advantages to their children. Although Bostrom and Sandberg (2009) are correct to argue that the risk of inequalities caused by CE is contingent upon its costs, the fundamental problem is if parents are willing to invest significant resources for interventions of only modest benefit, for example SAT preparation courses

[6] For example, a five-year combined undergraduate law degree with a Commonwealth contribution of $2089 per year.

(Buchmann, Condron, & Roscigno, 2010), one wonders what premium the well-to-do would offer when a one-point increase in IQ is associated with an increase in annual income of between $202 and $616 (Zagorsky, 2007).

Intelligence Enhancement as the Remedy

However, truly effective IE could just as effectively arrest inequalities as it could exacerbate them (Buchanan, Brock, Daniels, & Wikler, 2000). There is a certain irony in appealing to the Rawlsian theory of justice to argue that CE is inherently unjust (see Dubljevic, 2012) when Rawls delineates deep inequalities—structural in nature and inherited at birth—from shallow inequalities or those that arise later in life from voluntary choices (Arneson, 2008). Although the view that cognitive ability is the cause of socioeconomic inequality is a controversial one and beyond the scope of this chapter (see Herrnstein & Murray, 1994), the aspirations of IE are not antithetical to any account of justice that "nullifies the accidents of natural endowment" (Rawls, 1971, p. 15).

Under Rawls' Difference Principle, inequalities are justified as long as the least advantaged are better off than they would be under equal distribution (Rawls, 1971). Insofar as society is concerned with equality of opportunity, then it also ought to render them at the mercy of the natural lottery (Hauskeller, 2016). For example, intelligence mediates an individual's mobility to a different social class from the one than they have inherited from their parents (Nettle, 2003). Early intervention programmes such as Project Head Start have long sought the same outcomes as IE; the failure to do so has not come from a lack of trying, but observation that the benefit derived from these interventions are susceptible to the fadeout effect (see Protzko, 2015).

More broadly, however, there is also a general expectation that CE confers the same effects to all those who consume them, when there is empirical evidence to suggest there the effects of these agents are contingent upon baseline cognitive functioning (Cools & D'Esposito, 2011). Generally, an inverted U-shaped function is observed (see de Jongh, 2017). The dopaminergic psychostimulants, for example, have efficacy in improving working memory in the presence of hypofrontality; where there is none, however, they confer either no benefit or even deleterious effects on performance.

Intelligence Enhancement Meets Trickle-Down Theory

In their work standardising the national IQs of 192 nations against a normed "Greenwich IQ", Lynn and Vanhanen (2002, 2006, 2012) propose that increasing a nation's intelligence attenuates the income inequality observed within it. Although an inverse relationship has been identified between national IQ and measures of inequality, for example Gini coefficient (Meisenberg, 2012), just why this would be the case is not entirely clear. To date, the most cogent account links national IQ with a higher marginal tax rate (Kanazawa, 2009) or greater social transfers (Salahodjaev & Kanazawa, 2017). Although not a position that I endorse, one imagines a neoliberalism-inspired offshoot of trickle-down economics emerging to justify inequalities in IE on the basis that they will ultimately narrow inequalities.

Authenticity

It is commonly claimed that CE will render achievements derived from their use less deserved. Again, public sentiment seems to mirror these concerns (Faber, Savulescu, & Douglas, 2016; Partridge et al., 2014), suggesting that there is an overarching coherence to people's moral intuitions (de Sio, Faber, Savulescu, & Vincent, 2016). Central to this belief is the view that in order for our achievements to be truly deserved, we must be responsible for them. Thus, one cannot use CE without also ceding some of their agency over the outcome; if any praise is due, perhaps it ought to be directed to the apothecary instead (Sandel, 2007).

In rendering our achievements all that little more within our grasp, does CE undermine our achievements? If desert is contingent upon our responsibility for our actions, and insofar as they facilitate the same outcome with the need for less effort, then the use of any performance enhancer ought to render any achievement derived from them as being less deserved. Strictly speaking, however, all performance enhancers are deservedness attenuators, but oddly, they are hardly ever considered in the moral calculus (Bostrom & Sandberg, 2009; Harris, 2007).

The issue of authenticity raises an interesting question of whether CE somehow cheapens any accomplishments derived from their use. Those who claim so will tend to liken non-drug enhancements as the realisation of some latent potential, while any achievement derived from ligand-based enhancement is as hollow as buying a trophy. Rather than concern ourselves with whether CE makes people less deserving of praise, perhaps a more cogent account may emerge by considering how it is our beliefs on deservingness are shaped.

Causal Attributions of Responsibility

When I say your success is undeserved, I am making a causal claim about your responsibility over the outcome (Gerstenberg et al., 2018). Were it not for the presence of some external factor, there would be no achievement for us to speak of (Goodin, 1985). Beliefs regarding responsibility are largely predicated on the extent to which we believe that an individual's intentionality can be attributed to the outcome (Weiner, 1995). Consider the following scenario:

> Jared and Chet are identical twins who are preparing for their SATs. Jared is reasonably gifted due to several genetic polymorphisms that modulate prefrontal dopamine signalling. He gets good grades with only a moderate amount of effort. Chet, on the other hand, is not so lucky: despite sharing the same genes, he seems to have lost the (epi)genetic lottery and his potential flounders in a sea of methylated dopamine transporters. Unlike his brother, Chet has been preparing for his exams months in advance.

Suppose Jared and Chet were to score the same high grade in their exam, are they equally deserving of praise? Although they have both displayed agency over the outcome, responsibility is attributed to those who are perceived as having acted with intentionality; effort begets praise because it implies an agent has exerted volitional control over the outcome (Feather, 1992; Weiner, 1995). For the same achievement, we tend to view those who have exerted more effort as being more deserving, while those who have overcome adversity are lauded all that little bit more (Ogletree & Archer, 2012). In studying harder than his brother, most people would consider Chet more deserving of his mark.

Causal attributions, however, are only loosely based on reality because they are malleable to our own predilections and cognitive biases. We tend to attribute causality to a single large factor over many factors of smaller effect; proximal factors are given more weighing over distal ones; and, norm violating (Samland & Waldmann, 2016) or unusual (Gerstenberg et al., 2018) factors are selected over common ones. And quite predictably, when we are the actor, we attribute our achievements to our ability (Williams & Steffel, 2014), but of our failures, the last on our list tends to be ourselves (Mezulis, Abramson, Hyde, & Hankin, 2004).

Deservingness Is a Value Judgement

Alone, however, causal attributions cannot fully account for our judgements of deservedness. One can be viewed as responsible for their achievements, yet still be seen as undeserving of them (Feather, 1999):

> Chet prepared months in advance just to get the same mark as his brother Jared. For his next exam he adopts a different strategy. He buys methylphenidate (Ritalin) off a classmate with ADHD and takes it the same day as Jared starts study. As the drug courses through his brain, it increases the dopamine signalling in prefrontal cortex, offsetting his epigenetic disadvantage. Chet now has the same ability as Jared.

Suppose, for argument's sake, our twins were to yet again score an equally high mark in their exam, which of the two ought we to consider more deserving of praise? To be sure, with both brothers having the same ability, and having obtained the same mark, we would reason that they have both applied the same effort. If this is the case, then it stands to reason that they are equally responsible for their mark; Jared is no more responsible for any ability derived from his parent's genes, than Chet is for having the same genes whose epigenetic handicap has been lifted care of the Ritalin.

Yet something is amiss—most people would view Chet as less deserving of praise than his brother. He has broken several rules—legal and social. Some might consider him a drug cheat. Perhaps he has broken some honour code. How might we reconcile this schism between their equal degree of responsibility for their outcomes but disparate degrees of deservingness?

Work by Feather (1992, 2002) offers a coherent account for why this might be the case. Here, he frames judgements of deservingness as a function of a positively valued behaviour causing a positively valued outcome; individuals will only be viewed as deserving of their success if the outcome arises from positively regarded behaviour. Absent any regulatory framework, use of CE amounts to illicit drug use and necessarily precludes any successes derived from the use of these agents as deserved.

Other Considerations

I. Magnitude of Benefit Is Exaggerated

Public perceptions regarding the efficacy of CE appear to be heavily distorted and viewed under the existing paradigm of doping in sport (see Dodge et al., 2012). To date, the perceived improvement in performance that CE presumably offers has not been quantified. Achievements may be perceived as less deserved simply because counterfactual reasoning may attribute more of the outcome to an external cause.

II. Cognitive Enhancers May Be Effort Attenuators

Causal attribution of achievement outcomes is considered to be a function of ability, effort, task difficulty and luck, while deservingness is closely coupled with perceived effort (Leventhal & Michaels, 1971). Thus, by increasing cognitive ability, CE may lessen any attribution of achievement to the exertion of effort (but see Faber, Douglas, Heise, & Hewstone, 2015).

III. Universal Use Will Make Inauthentic Achievements Authentic Again

Causal inferences can only occur if there is covariance. Absent covariance, the ubiquitous adoption of CE would make inauthentic achievements authentic again, because it would not be possible to attribute the effects of CE to an individual's achievement.

Autonomy

While opinions diverge on whether CE ought to be embraced or eschewed, there is more or less widespread agreement in liberal democracies that an individual's decision to partake or not to partake in CE should be honoured (Greely et al., 2008). However, absent any direct coercion, we are still presented with the dilemma that CE may undermine the autonomy of those who do not wish to partake in its use (Dubljevic, 2012; Fukuyama, 2002). Although I have considered how IE may be causal in health outcomes that do not appear to impinge upon the well-being of others, insofar as IE would also improve functional outcomes for inherently positional goods (Harris, 2011), it nevertheless renders one person's benefit at the expense of another.

This has led to the legitimate concern that through the unregulated use of CE we may see the emergence of a "cognitive arms race" (Bostrom & Sandberg, 2009) where individuals may feel pressured to consume such agents simply to remain competitive. This indirect coercion is thought to influence the use of drugs in sport, or as Fukuyama (2002) correctly asks "if some move ahead, can anyone afford not to follow?" Public sentiment surrounding the use of psychostimulants seems to reflect an appreciation of the two competing interests, one where the individual ought to be free to, while the very use of these agents is thought to be driven by competitive pressures in the labour market (Forlini & Racine, 2009).

Coercion in the Classroom

Since the 1990s, there has been a significant growth in the rate of psychostimulant prescriptions, particularly among children. This has prompted concerns of diagnostic creep from minimal brain dysfunction in the 1960s to the current incarnation of ADHD (Graf, Miller, & Nagel, 2014). However, concerns regarding the coerced use of stimulants in schoolchildren is not a new phenomenon (Conrad, 1975). Following mostly inaccurate media reports in 1970 alleging 5% to 10% of schoolchildren in Omaha, Nebraska, were medicated with Ritalin after being identified by their teachers as having behaviour and learning problems, the ensuing outrage prompted a Congressional investigation into "Federal Involvement in the Use of Behavior Modification Drugs on Grammar School Children in the Right to Privacy Inquiry" (U.S. Congressional Report, 1970).

In 2001, however, a clear incentive to diagnose and treat students with ADHD when performance-based federal funding of public schools became codified in the No Child Left Behind Act. Due to variation in state-based performance requirements, there is evidence that states with more stringent accountability laws have higher rates of diagnosis of ADHD and treatment with psychostimulants (Bokhari & Schneider, 2011). Consistent with this view, it has also been found that a child's teacher is the most likely to first suggest a diagnosis of ADHD to the parent (Sax & Kautz, 2003). Countering concerns of coercion from school administrators, 14 states have passed "psychotropic medication laws" that prohibit public schools from recommending to parents that their child utilise psychostimulant or other drugs, to making the use of these medications a condition of enrolment (Fulton, Scheffler, & Hinshaw, 2015).

Coercion in the Workplace

Generally, coercion in the workplace has been framed as some form of overt attempt to pressure employees either to engage in CE or to make employment contingent upon such use (e.g. Appel, 2008). Under some circumstances, this form of coercion is entirely possible, but it hardly seems the type that we ought to concern ourselves with; it is coercion that is insidious that is the most difficult to arrest. Thus, Appel's (2008) proposal that employers be prohibited from discriminating based on one's use or non-use of CE is certainly correct, but ultimately an empty gesture. Given that intelligence is the best predictor of job performance (Schmidt & Hunter, 1998), it is hard to imagine why selection would be contingent upon the use of CE per se when the functional outcome of the enhancement is what is desired by an employer.

It is also likely that some industries will be more susceptible to CE than others. Although Dubljevic (2012) is correct to argue that long-haul trucking as an industry at risk of coercion to use psychostimulants, the wellspring from which coercion will likely emerge is from the replacement of entire industries through automation, the most susceptible being the transport and logistics industry (Clements & Kockelman, 2017). The need for less sleep will not be from the entrepreneur hoping to make an extra buck, but the sole breadwinner with mouths to feed and whose competitor's only sleep requirements are the few, odd minutes of a firmware update.

When Coercion Is Acceptable

Coercion may certainly be unpleasant, but under many circumstances it is well tolerated. Parents who denied access to social benefits unless they vaccinate their child are coerced into doing so. Likewise, water fluoridation and iodine fortification are therapeutic interventions that are imposed without informed consent.[7] The principle of autonomy is best honoured when we respect another's right to be irrational, foolish, or wrong, not when their wishes comport with our own. But while autonomy is highly valued by Western societies and is usually given priority against other principles, it is still considered *primus inter pares* in bioethics (Jennings, 2009). Thus, therapeutic coercion can be justified as follows:

[7] Although Schaefer, Kahane, and Savulescu (2014) argue that genetic CE would *increase* autonomy, it maps more faithfully to the coercion seen with iodine fortification.

> If the intervention is safe, it is reasonable to assume implied consent, and failing to administer the intervention in an intrinsically coercive manner would likely result in injustice, then therapeutic coercion is justified.

Suppose the IE drug we have considered was to ever come to fruition; would coercion of a type seen with vaccinations, water fluoridation, and iodine fortification be acceptable? I argue that it would and that doing so would be equally justified: I have never known someone wishing they were less intelligent, and choosing to forgo IE is no more rational than wishing to having poor dentition, a goitre's thyroid, or to be Patient Zero of a measles outbreak.

Even if IE were made freely available but entirely voluntary, intelligence so ubiquitously valued across cultures (Lippa, 2007) that prevailing mores would likely render IE inherently coercive. Any individual electing to forego IE would be viewed as foolhardy, while any parent opting their child out of the moral imperative to do so would be held in the same regard as a negligent parent.

References

Appel, J. M. (2008). When the boss turns pusher: A proposal for employee protections in the age of cosmetic neurology. *Journal of Medical Ethics, 34*, 616–618. https://doi.org/10.1136/jme.2007.022723

Arneson, R. (2008). Rawls, responsibility, and distributive justive. In M. Fleurbaey, M. Salles, & J. A. Weymark (Eds.), *Justice, political liberalism, and utilitarianism: Themes from Harsanyi and Rawls* (pp. 80–107). New York: Cambridge University Press.

Arria, A. M., Caldeira, K. M., Vincent, K. B., O'Grady, K. E., Cimini, M. D., Geisner, I. M., … Larimer, M. E. (2017). Do college students improve their grades by using prescription stimulants nonmedically? *Addictive Behaviors, 65*, 245–249. https://doi.org/10.1016/j.addbeh.2016.07.016

Arria, A. M., Caldeira, K. M., Vincent, K. B., O'Grady, K. E., & Wish, E. D. (2008). Perceived harmfulness predicts nonmedical use of prescription drugs among college students: Interactions with sensation-seeking. *Prevention Science, 9*, 191–201. https://doi.org/10.1007/s11121-008-0095-8

Banjo, O. C., Nadler, R., & Reiner, P. B. (2010). Physician attitudes towards pharmacological cognitive enhancement: Safety concerns are paramount. *PLoS ONE, 5*. https://doi.org/10.1371/journal.pone.0014322

Barmack, J. E. (1938). The effect of benzedrine sulfate (benzyl methyl carbinamine) upon the report of boredom and other factors. *Journal of Psychology: Interdisciplinary and Applied, 5*, 125–133. https://doi.org/10.1080/00223980.1938.9917557

Batty, G. D., Deary, I. J., Schoon, I., & Gale, C. R. (2007). Childhood mental ability in relation to cause-specific accidents in adulthood: The 1970 British Cohort Study. *QJM, 100*, 405–414. https://doi.org/10.1093/qjmed/hcm039

Batty, G. D., Deary, I. J., Tengstrom, A., & Rasmussen, F. (2008). IQ in early adulthood and later risk of death by homicide: Cohort study of 1 million men. *British Journal of Psychiatry, 193*, 461–465. https://doi.org/10.1192/bjp.bp.107.037424

Batty, G. D., Wennerstad, K. M., Smith, G. D., Gunnell, D., Deary, I. J., Tynelius, P., & Rasmussen, F. (2009). IQ in early adulthood and mortality by middle age. *Epidemiology, 20*, 100–109. https://doi.org/10.1097/EDE.0b013e31818ba076

Baweja, R., Mattison, R. E., & Waxmonsky, J. G. (2015). Impact of attention-deficit hyperactivity disorder on school performance: What are the effects of medication? *Pediatric Drugs, 17*, 459–477. https://doi.org/10.1007/s40272-015-0144-2

Board, C. (2017). College board. Retrieved from https://collegereadiness.collegeboard.org/pdf/understanding-sat-scores.pdf

Bokhari, F. A. S., & Schneider, H. (2011). School accountability laws and the consumption of psychostimulants. *Journal of Health Economics, 30*, 355–372. https://doi.org/10.1016/j.jhealeco.2011.01.007

Boring, E. G. (1923). Intelligence as the tests test it. *New Republic, 36*, 35–37.

Bostrom, N. (2008). Drugs can be used to treat more than disease. *Nature, 451*, 520–520. https://doi.org/10.1038/451520b

Bostrom, N., & Sandberg, A. (2009). Cognitive enhancement: Methods, ethics, regulatory challenges. *Science and Engineering Ethics, 15*, 311–341. https://doi.org/10.1007/s11948-009-9142-5

Bratsberg, B., & Rogeberg, O. (2017). Childhood socioeconomic status does not explain the IQ-mortality gradient. *Intelligence, 62*, 148–154. https://doi.org/10.1016/j.intell.2017.04.002

Buchanan, A., Brock, D. W., Daniels, N., & Wikler, D. (2000). *From chance to choice: Genetics and justice*. New York: Cambridge University Press.

Buchmann, C., Condron, D. J., & Roscigno, V. J. (2010). Shadow education, American style: Test preparation, the SAT and college enrollment. *Social Forces, 89*, 435–461. https://doi.org/10.1353/sof.2010.0105

Burleigh, T. J., Rubel, A. N., & Meegan, D. V. (2017). Wanting 'the whole loaf': Zero-sum thinking about love is associated with prejudice against consensual non-monogamists. *Psychology and Sexuality, 8*, 24–40. https://doi.org/10.1080/19419899.2016.1269020

Cakic, V. (2009). Smart drugs for cognitive enhancement: Ethical and pragmatic considerations in the era of cosmetic neurology. *Journal of Medical Ethics, 35*, 611–615. https://doi.org/10.1136/jme.2009.030882

Calvin, C. M., Batty, G. D., Der, G., Brett, C. E., Taylor, A., Pattie, A., … Deary, I. J. (2017). Childhood intelligence in relation to major causes of death in 68 year follow-up: Prospective population study. *BMJ, 2708*, j2708. https://doi.org/10.1136/bmj.j2708

Calvin, C. M., Batty, G. D., & Dreary, I. J. (2011). Cognitive epidemiology: Concepts, evidence, and future directions. In T. Chamorro-Premuzic, S. von Stumm, & A. Furnham (Eds.), *The Wiley-Blackwell handbook of individual differences* (pp. 427–460). London: Wiley-Blackwell.

Calvin, C. M., Deary, I. J., Fenton, C., Roberts, B. A., Der, G., Leckenby, N., & Batty, G. D. (2011). Intelligence in youth and all-cause-mortality: Systematic review with metaanalysis. *International Journal of Epidemiology, 40*, 626–644. https://doi.org/10.1093/ije/dyq190

Caviola, L., Mannino, A., Savulescu, J., & Faulmuller, N. (2014). Cognitive biases can affect moral intuitions about cognitive enhancement. *Frontiers in Systems Neuroscience, 8*, 1–5. https://doi.org/10.3389/fnsys.2014.00195

Chatterjee, A. (2007). Cosmetic neurology and cosmetic surgery: Parallels, predictions, and challenges. *Cambridge Quarterly of Healthcare Ethics, 16*, 129–137. https://doi.org/10.1017/S0963180107070156

Chen, E. S., & Tyler, T. R. (2001). Cloaking power: Legitimizing myths and the psychology of the advantaged. In A. Y. Lee-Chai & J. Bargh (Eds.), *The use and abuse of power: Multiple perspectives on the causes of corruption* (pp. 241–263). Philadelphia: Psychology Press.

Christensen, G. T., Mortensen, E. L., Christensen, K., & Osler, M. (2016). Intelligence in young adulthood and cause-specific mortality in the Danish Conscription Database – A cohort study of 728,160 men. *Intelligence, 59*, 64–71. https://doi.org/10.1016/j.intell.2016.08.001

Clements, L. M., & Kockelman, K. M. (2017). Economic effects of automated vehicles. *Transportation Research Record: Journal of the Transportation Research Board, 2606*(1), 106–114. https://doi.org/10.3141/2606-14

Conrad, P. (1975). The discovery of hyperkinesis: Notes on the medicalization of deviant behavior. *Social Problems, 23*, 12–21. https://doi.org/10.1525/sp.1975.23.1.03a00020

Cools, R., & D'Esposito, M. (2011). Inverted-U-shaped dopamine actions on human working memory and cognitive control. *Biological Psychiatry, 69*, e113–e125. https://doi.org/10.1016/j.biopsych.2011.03.028

Coomber, R. (2014). How social fear of drugs in the non-sporting world creates a framework for doping policy in the sporting world. *International Journal of Sport Policy and Politics, 6*, 171–193. https://doi.org/10.1080/19406940.2012.756824

Crisco, J. J., Fiore, R., Beckwith, J. G., Chu, J. J., Brolinson, G. P., Duma, S., … Greenwald, R. M. (2010). Frequency and location of head impact exposures in individual collegiate football players. *Journal of Athletic Training, 45*, 549–559. https://doi.org/10.4085/1062-6050-45.6.549

de Jongh, R. (2017). Overclocking the brain? The potential and limitations of cognitionenhancing drugs. In *Rethinking cognitive enhancement* (pp. 37–56). Oxford University Press. https://doi.org/10.1093/acprof:oso/9780198727392.003.0003

de Sio, F. S., Faber, N. S., Savulescu, J., & Vincent, N. A. (2016). Why less praise for enhanced performance? In F. Jotterand & V. Dubljevic (Eds.), *Cognitive enhancement: Ethical and policy implications in international perspectives* (pp. 27–41). New York: Oxford University Press. https://doi.org/10.1093/acprof:oso/9780199396818.003.0003

Dickson, H., Laurens, K. R., Cullen, A. E., & Hodgins, S. (2012). Meta-analyses of cognitive and motor function in youth aged 16 years and younger who subsequently develop schizophrenia. *Psychological Medicine, 42*, 743–755. https://doi.org/10.1017/S0033291711001693

Dodge, T., Williams, K. J., Marzell, M., & Turrisi, R. (2012). Judging cheaters: Is substance misuse viewed similarly in the athletic and academic domains? *Psychology of Addictive Behaviors, 26*, 678–682. https://doi.org/10.1037/a0027872

Donaldson, J. H., & Gray, M. (2012). Systematic review of grading practice: Is there evidence of grade inflation? *Nurse Education in Practice, 12*, 101–114. https://doi.org/10.1016/j.nepr.2011.10.007

Dubljevic, V. (2012). Principles of justice as the basis for public policy on psychopharmacological cognitive enhancement. *Law, Innovation and Technology, 4*, 67–83. https://doi.org/10.5235/175799612800650617

Dubljević, V., Sattler, S., & Racine, É. (2014). Cognitive enhancement and academic misconduct: A study exploring their frequency and relationship. *Ethics & Behavior, 24*, 408–420. https://doi.org/10.1080/10508422.2013.869747

Dunlop, M., & Savulescu, J. (2014). Distributive justice and cognitive enhancement in lower, normal intelligence. *Monash Bioethics Review, 32*, 189–204. https://doi.org/10.1007/s40592-014-0014-7

Editorial: Benzedrine sulfate "Pep Pills" [Editorial]. (1937). *Journal of the American Medical Association, 108*, 1973. https://doi.org/10.1001/jama.1937.02780230033012

Espenshade, T., & Chung, C. (2005). The opportunity cost of admission preference at elite universities. *Social Science Quarterly, 86*, 293–305.

Faber, N. S., Douglas, T., Heise, F., & Hewstone, M. (2015). Cognitive enhancement and motivation enhancement: An empirical comparison of intuitive judgments. *AJOB Neuroscience, 6*, 18–20. https://doi.org/10.1080/21507740.2014.991847

Faber, N. S., Savulescu, J., & Douglas, T. (2016). Why is cognitive enhancement deemed unacceptable? The role of fairness, deservingness, and hollow achievements. *Frontiers in Psychology, 7*, 1–12. https://doi.org/10.3389/fpsyg.2016.00232

Feather, N. T. (1992). An attributional and value analysis of deservingnes in success and failure situations. *British Journal of Social Psychology, 31*, 125–145. Retrieved from papers3://publication/uuid/6E6D9411-1078-4D1F-8BF1-C02FB00625D0.

Feather, N. T. (1999). Judgments of deservingness: Studies in the psychology of justice and achievement. *Personality and Social Psychology Review, 3*, 86–107. https://doi.org/10.1207/s15327957pspr0302_1

Feather, N. T. (2002). *Values, achievement, and justice: Studies in the psychology of deservingness*. New York: Kluwer Academic Publishers.

Finzi, E., & Rosenthal, N. E. (2016). Emotional proprioception: Treatment of depression with afferent facial feedback. *Journal of Psychiatric Research, 80*, 93–96. https://doi.org/10.1016/j.jpsychires.2016.06.009

Forlini, C., & Racine, E. (2009). Autonomy and coercion in academic "cognitive enhancement" using methylphenidate: Perspectives of key stakeholders. *Neuroethics, 2*, 163–177. https://doi.org/10.1007/s12152-009-9043-y

Forlini, C., & Racine, E. (2012). Stakeholder perspectives and reactions to "academic" cognitive enhancement: Unsuspected meaning of ambivalence and analogies. *Public Understanding of Science, 21*, 606–625. https://doi.org/10.1177/0963662510385062

Froehlich, T. E., Lanphear, B. P., Epstein, J. N., Barbaresi, W. J., Katusic, S. K., & Kahn, R. S. (2007). Prevalence, recognition, and treatment of attention-deficit/hyperactivity disorder in a national sample of US children. *Archives of Pediatrics & Adolescent Medicine, 161*, 857–864.

Fukuyama, F. (2002). *Our posthuman future: Consequences of the biotechnology revolution.* New York: Farrar, Straus and Giroux.

Fulton, B. D., Scheffler, R. M., & Hinshaw, S. P. (2015). State variation in increased ADHD prevalence: Links to NCLB school accountability and state medication laws. *Psychiatric Services, 66*, 1074–1082. https://doi.org/10.1176/appi.ps.201400145

Gale, C. R., Batty, G. D., Mcintosh, A. M., Porteous, D. J., Deary, I. J., & Rasmussen, F. (2013). Is bipolar disorder more common in highly intelligent people A cohort study of a million men. *Molecular Psychiatry, 18*, 190–194. https://doi.org/10.1038/mp.2012.26

Gale, C. R., Batty, G. D., Tynelius, P., Deary, I. J., & Rasmussen, F. (2010). Intelligence in early adulthood and subsequent hospitalization for mental disorders. *Epidemiology, 21*, 70–77. https://doi.org/10.1097/EDE.0b013e3181c17da8

Gerstenberg, T., Ullman, T. D., Nagel, J., Kleiman-Weiner, M., Lagnado, D. A., & Tenenbaum, J. B. (2018). Lucky or clever? From expectations to responsibility judgments. *Cognition, 177*, 122–141. https://doi.org/10.1016/j.cognition.2018.03.019

Goodin, R. E. (1985). Vulnerabilities and responsibilities: An ethical defense of the welfare state. *American Political Science Review, 79*(3), 775–87.

Goodman, R. (2010). Cognitive enhancement, cheating, and accomplishment. *Kennedy Institute of Ethics Journal, 20*, 145–160. https://doi.org/10.1353/ken.0.0309

Gottfredson, L. S. (2003). G, jobs and life. In H. Nyborg (Ed.), *The scientific study of general intelligence: A tribute to Arthur Jensen* (pp. 293–342). Pergamon.

Gottfredson, L. S. (2004). Intelligence: Is it the epidemiologists' elusive "fundamental cause" of social class inequalities in health? *Journal of Personality and Social Psychology, 86*, 174–199. https://doi.org/10.1037/0022-3514.86.1.174

Graf, W. D., Miller, G., & Nagel, S. K. (2014). Addressing the problem of ADHD medication as neuroenhancements. *Expert Review of Neurotherapeutics, 14*, 569–581. https://doi.org/10.1586/14737175.2014.908707

Greely, H., Sahakian, B., Harris, J., Kessler, R. C., Gazzaniga, M., Campbell, P., & Farah, M. J. (2008). Towards responsible use of cognitive-enhancing drugs by the healthy. *Nature, 456*, 702–705. https://doi.org/10.1038/456702a

Green, S. P. (2004). Cheating. *Law and Philosophy, 23*, 137–185. https://doi.org/10.1023/B:LAPH.0000011918.29196.ec

Gyngell, C., & Selgelid, M. J. (2016). Human enhancement: Conceptual clarity and moral significance. In *The ethics of human enhancement* (pp. 111–126). Oxford: Oxford University Press. https://doi.org/10.1093/acprof:oso/9780198754855.003.0008

Harris, J. (2007). *Enhancing evolution: The ethical case for making better people.* Princeton, NJ: Princeton University Press.

Harris, J. (2011). Chemical cognitive enhancement: Is it unfair, unjust, discriminatory, or cheating for healthy adults to use smart drugs? In J. Illes (Ed.), *Oxford handbook of neuroethics* (1st ed., pp. 265–272). Oxford: Oxford University Press.

Hart, C. L., Taylor, M. D., Whalley, L. J., Starr, J. M., Hole, D. J., Wilson, V., & Deary, I. J. (2003). Childhood IQ, social class, deprivation and their relationship with mortality and morbidity risk in later life: Prospective observational study linking the Scottish Mental Survey 1932 and the Midspan studies. *Psychosomatic Medicine, 65*, 877–883. https://doi.org/10.1097/01.PSY.0000088584.82822.86

Hauskeller, M. (2016). Levelling the playing field. In *The ethics of human enhancement* (pp. 198–210). New York: Oxford University Press. https://doi.org/10.1093/acprof:oso/9780198754855.003.0014

Herrnstein, R. J., & Murray, C. (1994). *The bell curve: Intelligence and class structure in American life*. New York: Simon & Schuster.

Hupli, A. (2013). *Perceptions, practices and ethics of (non)-prescription cognitive enhancement drugs: A case study among academic youth in Amsterdam*. Master's thesis, University of Amsterdam.

Hurwitz, M. (2011). The impact of legacy status on undergraduate admissions at elite colleges and universities. *Economics of Education Review, 30*, 480–492. https://doi.org/10.1016/j.econedurev.2010.12.002

Ilieva, I., Boland, J., & Farah, M. J. (2013). Objective and subjective cognitive enhancing effects of mixed amphetamine salts in healthy people. *Neuropharmacology, 64*, 496–505. https://doi.org/10.1016/j.neuropharm.2012.07.021

Ilieva, I. P., & Farah, M. J. (2013). Enhancement stimulants: Perceived motivational and cognitive advantages. *Frontiers in Neuroscience, 7*, 1–6. https://doi.org/10.3389/fnins.2013.00198

Jaeggi, S. M., Buschkuehl, M., Jonides, J., & Perrig, W. J. (2008). Improving fluid intelligence with training on working memory. *Proceedings of the National Academy of Sciences of the United States of America, 105*, 6829–6833. https://doi.org/10.1073/pnas.0801268105

Jennings, B. (2009). *Autonomy* (B. Steinbock, Ed.). New York: Oxford University Press. https://doi.org/10.1093/oxfordhb/9780199562411.003.0004

Jensen, A. R. (1998). *The G factor: The science of mental ability*. Westport, CT: Praeger.

Jerrim, J., Chmielewski, A. K., & Parker, P. (2015). Socioeconomic inequality in access to high-status colleges: A cross-country comparison. *Research in Social Stratification and Mobility, 42*, 20–32. https://doi.org/10.1016/j.rssm.2015.06.003

Johnson, V. E. (2003). *Grade inflation*. New York: Springer. https://doi.org/10.1007/b97309

Kanazawa, S. (2009). IQ and the values of nations. *Journal of Biosocial Science, 41*, 537. https://doi.org/10.1017/S0021932009003368

Kass, L. (2003). *Beyond Therapy: Biotechnology and the Pursuit of Happiness*. New York: Harper Collins.

King, M. D., Jennings, J., & Fletcher, J. M. (2014). Medical adaptation to academic pressure. *American Sociological Review, 79*, 1039–1066. https://doi.org/10.1177/0003122414553657

Kramer, T., Irmak, C., Block, L. G., & Ilyuk, V. (2012). The effect of a no-pain, no-gain lay theory on product efficacy perceptions. *Marketing Letters, 23*, 517–529. https://doi.org/10.1007/s11002-012-9165-6

Lager, E., Melin, B., Hemmingsson, T., & Sörberg Wallin, A. (2017). The evolving relationship between premorbid intelligence and serious depression across the lifespan – A longitudinal study of 43,540 Swedish men. *Journal of Affective Disorders, 211*, 37–43. https://doi.org/10.1016/j.jad.2016.12.051

Lavazza, A. (2016). A Rawlsian version of the opportunity maintenance thesis. *The American Journal of Bioethics, 16*, 50–52. https://doi.org/10.1080/15265161.2016.1170236

Leventhal, G. S., & Michaels, J. W. (1971). Locus of cause and equity motivation as determinants of reward allocation. *Journal of Personality and Social Psychology, 17*, 229–235. https://doi.org/10.1037/h0030602

Liakoni, E., Schaub, M. P., Maier, L. J., Glauser, G. V., & Liechti, M. E. (2015). The use of prescription drugs, recreational drugs, and "soft enhancers" for cognitive enhancement among Swiss secondary school students. *PLoS ONE, 10*, 1–12. https://doi.org/10.1371/journal.pone.0141289

Lindgren, M., Eriksson, P., Rosengren, A., Robertson, J., Schiöler, L., Schaufelberger, M., … Åberg, M. (2018). Cognitive performance in late adolescence and long-term risk of early heart failure in Swedish men methods and results. *European Journal of Heart Failure,* 4–6. https://doi.org/10.1002/ejhf.1163

Lippa, R. A. (2007). The preferred traits of mates in a cross-national study of heterosexual and homosexual men and women: An examination of biological and cultural influences. *Archives of Sexual Behavior, 36*, 193–208. https://doi.org/10.1007/s10508006-9151-2

Lynch, N., & Berry, D. (2007). Differences in perceived risks and benefits of herbal, over-the-counter conventional, and prescribed conventional, medicines, and the implications of this for the safe and effective use of herbal products. *Complementary Therapies in Medicine, 15*, 84–91. https://doi.org/10.1016/j.ctim.2006.06.007

Lynn, R., & Vanhanen, T. (2002). *IQ and the wealth of nations.* Westport, CT: Praeger.

Lynn, R., & Vanhanen, T. (2006). *IQ and global inequality.* Washington Summit Publishers.

Lynn, R., & Vanhanen, T. (2012). National IQs: A review of their educational, cognitive, economic, political, demographic, sociological, epidemiological, geographic and climatic correlates. *Intelligence, 40*, 226–234. https://doi.org/10.1016/j.intell.2011.11.004

Magid, M., Finzi, E., Kruger, T. H. C., Robertson, H. T., Keeling, B. H., Jung, S., … Wollmer, M. A. (2015). Treating depression with botulinum toxin: A pooled analysis of randomized controlled trials. *Pharmacopsychiatry, 48*, 205–210. https://doi.org/10.1055/s-0035-1559621

Martin, L. T., & Kubzansky, L. D. (2005). Childhood cognitive performance and risk of mortality: A prospective cohort study of gifted individuals. *American Journal of Epidemiology, 162*, 887–890. https://doi.org/10.1093/aje/kwi300

Meegan, D. V. (2010). Zero-sum bias: Perceived competition despite unlimited resources. *Frontiers in Psychology, 1*, 1–7. https://doi.org/10.3389/fpsyg.2010.00191

Meisenberg, G. (2012). National IQ and economic outcomes. *Personality and Individual Differences, 53*, 103–107. https://doi.org/10.1016/j.paid.2011.06.022

Melby-Lervåg, M., Redick, T. S., & Hulme, C. (2016). Working memory training does not improve performance on measures of intelligence or other measures of "far transfer": Evidence from a meta-analytic review. *Perspectives on Psychological Science, 11*, 512–534. https://doi.org/10.1177/1745691616635612

Merriman, S. J. W. (1848). *Arguments against the indiscriminate use of chloroform in midwifery*. Retrieved from https://www.woodlibrarymuseum.org/library/pdf/S_ADHT.pdf

Mez, J., Daneshvar, D. H., Kiernan, P. T., Abdolmohammadi, B., Alvarez, V. E., Huber, B. R., … McKee, A. C. (2017). Clinicopathological evaluation of chronic traumatic encephalopathy in players of American football. *JAMA, 318*, 360–370. https://doi.org/10.1001/jama.2017.8334

Mezulis, A. H., Abramson, L. Y., Hyde, J. S., & Hankin, B. L. (2004). Is there a universal positivity bias in attributions? A meta-analytic review of individual, developmental, and cultural differences in the self-serving attributional bias. *Psychological Bulletin, 130*, 711–747. https://doi.org/10.1037/0033-2909.130.5.711

Moynihan, R., Heath, I., & Henry, D. (2002). Selling sickness: The pharmaceutical industry and disease mongering. *BMJ, 324*, 886–890. https://doi.org/10.1136/bmj.324.7342.886

Neale, C., Camfield, D., Reay, J., Stough, C., & Scholey, A. (2013). Cognitive effects of two nutraceuticals Ginseng and Bacopa benchmarked against modafinil: A review and comparison of effect sizes. *British Journal of Clinical Pharmacology, 75*, 728–737. https://doi.org/10.1111/bcp.12002

Nettle, D. (2003). Intelligence and class mobility in the British population. *British Journal of Psychology, 94*, 551–561. https://doi.org/10.1348/000712603322503097

New York Times. (1937, April 10). Efficiency of brain held due to its 'fuel'; cells found speeded up by synthetic drug. *The New York Times*, p. 6.

Norton, M. I., & Sommers, S. R. (2011). Whites see racism as a zero-sum game that they are now losing. *Perspectives on Psychological Science, 6*, 215–218. https://doi.org/10.1177/1745691611406922

Nutt, D. (2009). Equasy – An overlooked addiction with implications for the current debate on drug harms. *Journal of Psychopharmacology, 23*, 3–5. https://doi.org/10.1177/0269881108099672

O'Toole, B. I. (1990). Intelligence and behaviour and motor vehicle accident mortality. *Accident Analysis & Prevention, 22*, 211–221. https://doi.org/10.1016/00014575(90)90013-B

Ogletree, S. M., & Archer, R. L. (2012). Pats on the back or pointing the finger: Judgments of praise and blame. *E-Journal of Applied Psychology, 8*. https://doi.org/10.7790/ejap.v8i1.345

Osler, M., Christensen, G. T., Garde, E., Mortensen, E. L., & Christensen, K. (2017). Cognitive ability in young adulthood and risk of dementia in a cohort of Danish men, brothers, and twins. *Alzheimer's and Dementia*, 1–9. https://doi.org/10.1016/j.jalz.2017.04.003

Ott, M., Bozeman, B., & Taggart, G. (2018). Risks and rewards of college football: Who would accept a scholarship knowing the chances of physical harm? *Social Science Quarterly*. https://doi.org/10.1111/ssqu.12491

Parens, E. (1998). Special supplement: Is better always good? The enhancement project. *The Hastings Center Report, 28*, S1–S17. https://doi.org/10.2307/3527981

Partridge, B., Lucke, J., & Hall, W. (2014). "If you're healthy you don't need drugs": Public attitudes towards "brain doping" in the classroom and "legalised doping" in sport. *Performance Enhancement and Health, 3*, 20–25. https://doi.org/10.1016/j.peh.2014.03.001

Pickering, G. W. (1968). *High blood pressure* (2nd ed.). London: Churchill.

President's Council on Bioethics. (2003). *Beyond therapy: Biotechnology and the pursuit of happiness.*

Protzko, J. (2015). The environment in raising early intelligence: A meta-analysis of the fadeout effect. *Intelligence, 53*, 202–210. https://doi.org/10.1016/j.intell.2015.10.006

Quednow, B. B. (2010). Ethics of neuroenhancement: A phantom debate. *BioSocieties, 5*, 153–156. https://doi.org/10.1057/biosoc.2009.13

Rawls, J. (1971). *A theory of justice*. Harvard University Press.

Rose, S. (2005). *The future of the brain: The promise and perils of tomorrow's neuroscience.* New York: Oxford University Press.

Rothstein, W. G. (2003). *Public health and the risk factor: A history of an uneven medical revolution*. Rochester, NY: University of Rochester Press.

Rousseau, J.-J. (1992). *Discourse on the Origin of Inequality*. (D.A. Cress, Trans.). Indianapolis, IN: Hackett Publishing Co. (Original work published 1754).

Salahodjaev, R., & Kanazawa, S. (2017). Why do societies with higher average cognitive ability have lower income inequality? The role of redistributive policies. *Journal of Biosocial Science*, 1–18. https://doi.org/10.1017/S0021932017000268

Samland, J., & Waldmann, M. R. (2016). How prescriptive norms influence causal inferences. *Cognition, 156*, 164–176. https://doi.org/10.1016/j.cognition.2016.07.007

Sandel, M. J. (2007). *The case against perfection: Ethics in the age of genetic engineering*. Cambridge, MA: Harvard University Press.

Santoni de Sio, F., Robichaud, P., & Vincent, N. A. (2014). Who should enhance? Conceptual and normative dimensions of cognitive enhancement. *Humana Mente Journal of Philosophical Studies, 26*, 179–197.

Savulescu, J. (2005). New breeds of humans: The moral obligation to enhance. *Reproductive BioMedicine Online, 10*, 36–39. https://doi.org/10.1016/S1472-6483(10)62202-X

Savulescu, J. (2006). Justice, fairness, and enhancement. *Annals of the New York Academy of Sciences, 1093*, 321–338. https://doi.org/10.1196/annals.1382.021

Savulescu, J., Foddy, B., & Clayton, M. (2005). Enhancing drugs in sport. *British Journal of Sports Medicine*, 666–670. https://doi.org/10.1136/bjsm.2003.005249

Sax, L., & Kautz, K. J. (2003). Who first suggests the diagnosis of attention deficit/hyperactivity disorder? *Annals of Family Medicine, 1*, 171–174. https://doi.org/10.1370/afm.3

Schaefer, G. O., Kahane, G., & Savulescu, J. (2014). Autonomy and enhancement. *Neuroethics, 7*, 123–136. https://doi.org/10.1007/s12152-013-9189-5

Schelle, K. J., Faulmüler, N., Caviola, L., & Hewstone, M. (2014). Attitudes toward pharmacological cognitive enhancement: A review. *Frontiers in Systems Neuroscience, 8*, 1–14. https://doi.org/10.3389/fnsys.2014.00053

Schermer, M. (2008). On the argument that enhancement is "cheating". *Journal of Medical Ethics, 34*, 85–88. https://doi.org/10.1136/jme.2006.019646

Schermer, M., & Bolt, I. (2011). What's in a name? ADHD and the gray area between treatment and enhancement. In J. Savulescu, R. ter Meulen, & G. Kahane (Eds.), *Enhancing human capacities* (pp. 179–193). Oxford: Blackwell Publishing Ltd.. https://doi.org/10.1002/9781444393552.ch12

Scheske, C., & Schnall, S. (2012). The ethics of "smart drugs": Moral judgments about healthy people's use of cognitive-enhancing drugs. *Basic and Applied Social Psychology, 34*, 508–515. https://doi.org/10.1080/01973533.2012.711692

Schleim, S., & Quednow, B. B. (2017). Debunking the ethical neuroenhancement debate. In R. ter Meulen, A. Mohammed, & W. Hall (Eds.), *Rethinking cognitive enhancement* (pp. 164–176). Oxford University Press. https://doi.org/10.1093/acprof:oso/9780198727392.003.0010

Schmidt, F. L., & Hunter, J. E. (1998). The validity and utility of selection methods in personnel psychology: Practical and theoretical implications of 85 years of research findings. *Psychological Bulletin, 124*, 262–274. https://doi.org/10.1037/00332909.124.2.262

Shadowen, S. D., Tulante, S. P., & Alpern, S. L. (2009). No distinctions except those which merit originates: The unlawfulness of legacy preferences in public and private universities. *Santa Clara Law Review, 49*, 51–136.

Simoni, Z. R., & Drentea, P. (2016). ADHD, socioeconomic status, medication use, and academic ethic. *Sociological Focus, 49*, 119–132. https://doi.org/10.1080/00380237.2016.1107713

Sirola, N., & Pitesa, M. (2017). Economic downturns undermine workplace helping by promoting a zero-sum construal of success. *Academy of Management Journal, 60*, 1339–1359. https://doi.org/10.5465/amj.2015.0804

Spriggs, M. (2005). Hypoxic air machines: Performance enhancement through effective training – Or cheating? *Journal of Medical Ethics, 31*, 112–113. https://doi.org/10.1136/jme.2003.005280

Teter, C. J., DiRaimo, C. G., West, B. T., Schepis, T. S., & McCabe, S. E. (2018). Nonmedical use of prescription stimulants among US high school students to help study: Results from a national survey. *Journal of Pharmacy Practice.* https://doi.org/10.1177/0897190018783887

Time. (1937, May). Pep-pill poisoning. *Time*. Retrieved from https://content.time.com/time/magazine/article/0,9171,757775,00.html

Trahan, L. H., Stuebing, K. K., Fletcher, J. M., & Hiscock, M. (2014). The Flynn effect: A meta-analysis. *Psychological Bulletin, 140*, 1332–1360. https://doi.org/10.1037/a0037173

United States Congressional House Government and Operations Committee. (1970). *Federal involvement in the use of behavior modificaton drugs on grammar school children in the right to privacy inquiry*. Washington, DC: U.S. Government Printing Office.

Urfer-Parnas, A., Lykke Mortensen, E., Sæbye, D., & Parnas, J. (2010). Premorbid IQ in mental disorders: A Danish draft-board study of 7486 psychiatric patients. *Psychological Medicine, 40*, 547. https://doi.org/10.1017/S0033291709990754

Volkow, N. D., Wang, G., Fowler, J. S., Telang, F., Maynard, L., Logan, J., … Swanson, J. M. (2004). Evidence that methylphenidate enhances the saliency of a mathematical task by increasing dopamine in the human brain. *American Journal of Psychiatry, 161*, 1173–1180. https://doi.org/10.1176/appi.ajp.161.7.1173

Vrecko, S. (2013). Just how cognitive is "cognitive enhancement"? On the significance of emotions in university students' experiences with study drugs. *AJOB Neuroscience, 4*, 4–12. https://doi.org/10.1080/21507740.2012.740141

Weiner, B. (1995). *Judgments of responsibility: A foundation for a theory of social conduct*. New York: Guilford Press.

Wellman, J. D., Liu, X., & Wilkins, C. L. (2016). Priming status-legitimizing beliefs: Examining the impact on perceived anti-White bias, zero-sum beliefs, and support for affirmative action among White people. *The British Journal of Social Psychology, 55*, 426–437. https://doi.org/10.1111/bjso.12133

Whitley, E., Batty, G. D., Gale, C. R., Deary, I. J., Tynelius, P., & Rasmussen, F. (2010a). Intelligence in early adulthood and subsequent risk of assault: Cohort study of 1,120,998 Swedish men. *Psychosomatic Medicine, 72*, 390–396. https://doi.org/10.1097/PSY.0b013e3181d137e9

Whitley, E., Batty, G. D., Gale, C. R., Deary, I. J., Tynelius, P., & Rasmussen, F. (2010b). Intelligence in early adulthood and subsequent risk of unintentional injury over two decades: Cohort study of 1,109,475 Swedish men. *Journal of Epidemiology & Community Health, 64*, 419–425. https://doi.org/10.1136/jech.2009.100669

Williams, E. F., & Steffel, M. (2014). Double standards in the use of enhancing products by self and others. *Journal of Consumer Research, 41*, 506–525. https://doi.org/10.1086/676750

Wrulich, M., Stadler, G., Brunner, M., Keller, U., & Martin, R. (2015). Childhood intelligence predicts premature mortality: Results from a 40-year population-based longitudinal study. *Journal of Research in Personality, 58*, 6–10. https://doi.org/10.1016/j.jrp.2015.06.003

Young, M. E., Norman, G. R., & Humphreys, K. R. (2008). The role of medical language in changing public perceptions of illness. *PLoS ONE, 3*, 1–6. https://doi.org/10.1371/journal.pone.0003875

Zagorsky, J. L. (2007). Do you have to be smart to be rich? The impact of IQ on wealth, income and financial distress. *Intelligence, 35*, 489–501. https://doi.org/10.1016/j.intell.2007.02.003

Zammit, S., Allebeck, P., David, A. S., Dalman, C., Hemmingsson, T., Lundberg, I., & Lewis, G. (2004). A longitudinal study of premorbid IQ score and risk of developing schizophrenia, bipolar disorder, severe depression, and other nonaffective psychoses. *Archives of General Psychiatry, 61*, 354. https://doi.org/10.1001/archpsyc.61.4.354

Zimmermann, M. B. (2012). The effects of iodine deficiency in pregnancy and infancy. *Paediatric and Perinatal Epidemiology, 26*, 108–117. https://doi.org/10.1111/j.13653016.2012.01275.x

8

Afterword

Jamie L. Tully, Mark Forshaw, Catharine Montgomery, and Matthew Hall

Overview

This edited collection has discussed three broad themes: the uses and effects of nutraceuticals, the uses and effects of pharmacological substances like pharmacological cognitive enhancer (PCE), and the broad ethical implications of PCE use. Each chapter has been self-contained, but here the various topics are brought together to discuss how cognitive enhancement (CE) drugs might further be explored, including the possible future directions for research, the ethical and moral implications of use, and any potential for physical harm. Exploring CE use is a developing field of study, and much of the pre-existing research examining the

J. L. Tully • M. Forshaw • C. Montgomery
Liverpool John Moores University, Liverpool, UK
e-mail: J.Tully@2010.ljmu.ac.uk; M.J.Forshaw@ljmu.ac.uk; C.A.Montgomery@ljmu.ac.uk

M. Hall (✉)
Arden University, Coventry, UK

University of Derby, Derby, UK
e-mail: mhall@arden.ac.uk

© The Author(s) 2021
M. Hall et al. (eds.), *Chemically Modified Minds*,
https://doi.org/10.1007/978-981-15-6771-1_8

various substances used to achieve CE is over-reliant on cross-sectional data and acute administration research. Consequently, high-quality data examining the impact and epidemiology of long-term use is extremely limited and leaves much to speculation. Nevertheless, by examining drugs with similar pharmacological features to PCE, such as amphetamine and cocaine, we might better understand the dangers associated with repeated use of these substances. Furthermore, many drugs covered in this edited collection can now be bought online despite pre-existing restrictions under UK law; as a consequence, there is an imminent need for a stronger understanding of CE use.

Potential Harms

Nutraceutical substances are organic plant extracts lauded for their health benefits (Ward et al., 2019); as such, the physical harms associated with their use are minimal as bioactive agents are known to increase physical well-being over time (Zeisel, 1999). Some relatively mild adverse side effects have been recorded with single use, including headaches, nausea and vomiting (Ronis, Pedersen, & Watt, 2018). However, hidden dangers exist from the online market, where it is feared that long-term unregulated production by unsupervised retailers could lead to harm for the consumer, particularly if products being consumed are counterfeit (Daud, Jalil, Azmi, Ismail, & Safuan, 2017). Dangers associated with counterfeit products appear self-evident, as some contaminated drugs have been found to contain scheduled poisons and illegal substances (Pin, 2013). Long-term use of counterfeits could therefore have an untold impact on health, making it important to purchase these drugs from licensed speciality retailers. Similar concerns exist with PCE, as there is a growing market for online sales of these drugs to get around prescription-only restrictions, and in the UK, large quantities of modafinil and methylphenidate are purchasable which exceed what would typically be prescribed, increasing the risk of misuse (Dursun, Dunn, & McKay, 2019; Hockenhull, Wood, & Dargan, 2020). Nevertheless, common side effects reported with single use of these substances are mild and include insomnia, nausea, tachycardia and headaches (Caldwell, Caldwell, Smith, &

Brown, 2004; Wesensten, Killgore, & Balkin, 2005). Furthermore, neurochemical similarities with illegal stimulants raise concerns about the potential neurotoxicity of continual PCE use, but a lack of data assessing long-term daily dosing (>3 months), particularly with modafinil, means that the possible risks are largely unknown.

Ethical Considerations

The ethical impact of CE use is a key theme in this edited collection. Many of the substances described in this book have been used to meet certain study and work-based demands. Perceived benefits to behavioural and cognitive performance are a driving factor in the use of CE at university (Maier, Liakoni, Schildmann, Schaub, & Liechti, 2015), but any educational advantages derived from use could be viewed as unfair. If grades are improved as a result of CE use, then the legitimacy of the achievement comes into question. For instance, two-thirds (68%) of Swiss university students surveyed indicated that academic performance achieved with PCE was seen as less worthy of recognition, and the vast majority (80%) felt that it was morally unacceptable (Maier et al., 2015). Universities must, therefore, make policy decisions concerning CE use in academia, to decide if these drugs provide an unfair advantage. Moreover, if prevalence continues to rise at university, then CE use might become normalised, and students could be pressured to engage with these techniques, creating educational inequality with those who do not. Universities must, therefore, take a balanced position on CE use and attempt to further understand the ethical implications of rising prevalence.

Future Directions

The potential risks associated with many of the substances discussed in this collection reveal several directions for future research. Pre-existing studies with pharmaceutical stimulants primarily focus on adolescent ADHD sufferers (Advokat & Scheithauer, 2013), but little high-quality data exists with healthy people using these drugs as part of a long-term

PCE strategy. This is especially true with modafinil, which recent data reveals to be the most popular PCE among UK university students (Maier, Ferris, & Winstock, 2018). Nevertheless, at the time of writing, no studies have been identified which assess long-term (>3 months) modafinil use in terms of the drug's impact on cognition, behaviour and health. Future research should examine this substance and other PCE (methylphenidate and d-amphetamine) further, to gain a better insight into whether or not these drugs adversely affect health. Studies should also consider ethics and morality and, similarly to Maier et al. (2015) with their Swiss cohort, investigate how opinions on use (i.e., the moral acceptability) can influence CE behaviour in the UK. Such research might enable universities to better understand student attitudes to these drugs and how they can impact engagement with these substances.

Summary

Further research is required to advance the understanding of CE use in the UK. This book has revealed some benefits linked to consumption of these drugs, including health benefits with nutraceuticals and some modest cognitive and behavioural enhancements with PCE. Nonetheless, issues relating to harm when dosing regimens are unregulated or counterfeit products are purchased online remain a concern, as do neurochemical similarities with PCE to illegal stimulants. The ethical implications of the rising popularity of these drugs must also be addressed, and universities must take a clear policy position on these substances and provide balanced information to their students regarding use and potential dangers. This edited collection has opened up the conversation surrounding CE drugs, and it is our hope that these questions will not remain unanswered.

References

Advokat, C. D., & Scheithauer, M. (2013). Attention-deficit hyperactivity disorder (ADHD) stimulant medications as cognitive enhancers. *Frontiers in Neuroscience, 7*, 82.

Caldwell, J. A., Caldwell, J. L., Smith, J. K., & Brown, D. L. (2004). Modafinil's effects on simulator performance and mood in pilots during 37 h without sleep. *Aviation, Space, and Environmental Medicine, 75*(9), 777–784.

Daud, M., Jalil, J. A., Azmi, I. M. A. G., Ismail, S. F., & Safuan, S. (2017). *'Unsafe' nutraceuticals products on the Internet: The need for stricter regulation in Malaysia.* Paper presented at the 2017 5th International Conference on Cyber and IT Service Management (CITSM).

Dursun, S., Dunn, M., & McKay, F. H. (2019). The availability and acquisition of modafinil on the internet. *Drug and Alcohol Review, 38*(6), 699–702.

Hockenhull, J., Wood, D. M., & Dargan, P. I. (2020). The availability of modafinil and methylphenidate purchased from the Internet in the United Kingdom without a prescription. *Substance Use & Misuse, 55*(1), 56–65.

Maier, L. J., Ferris, J. A., & Winstock, A. R. (2018). Pharmacological cognitive enhancement among non-ADHD individuals—A cross-sectional study in 15 countries. *International Journal of Drug Policy, 58*, 104–112.

Maier, L. J., Liakoni, E., Schildmann, J., Schaub, M. P., & Liechti, M. E. (2015). Swiss university students' attitudes toward pharmacological cognitive enhancement. *PloS One, 10*(12), e0144402.

Pin, C. H. (2013). Produk makanan berasaskan herba dan keperluan jaminan. *Buletin Teknologi, 3*, 79–86.

Ronis, M. J., Pedersen, K. B., & Watt, J. (2018). Adverse effects of nutraceuticals and dietary supplements. *Annual Review of Pharmacology and Toxicology, 58*, 583–601.

Ward, W. E., Chilibeck, P. D., Comelli, E. M., Duncan, A. M., Phillips, S. M., Robinson, L. E., & Stellingwerff, T. (2019). Research in nutritional supplements and nutraceuticals for health, physical activity, and performance: Moving forward. *Applied Physiology, Nutrition, and Metabolism, 44*(5), 455–460.

Wesensten, N. J., Killgore, W. D., & Balkin, T. J. (2005). Performance and alertness effects of caffeine, dextroamphetamine, and modafinil during sleep deprivation. *Journal of Sleep Research, 14*(3), 255–266.

Zeisel, S. H. (1999). Regulation of "nutraceuticals". *American Association for the Advancement of Science, 285*(5435), 1853–1855.

Index

© The Author(s) 2021
M. Hall et al. (eds.), *Chemically Modified Minds*,
https://doi.org/10.1007/978-981-15-6771-1

The manufacturer's authorised representative in the EU is Springer Nature Customer Service Centre GmbH, Europaplatz 3, 69115 Heidelberg, Germany. If you have any concerns regarding our products, please contact ProductSafety@springernature.com

Printed and bound by CPI Group (UK) Ltd, Croydon, CR0 4YY
30/09/2026
02225457-0002